OpenStack
構築手順書 [Kilo版]

日本仮想化技術株式会社 =著

購入者限定
無料FAQ
サポート付

OSSクラウド環境基盤の大本命！
<Ubuntu Server 14.04 ベース>
Keystone / Glance / Nova / Neutron / Cinder

インプレス

- 本書は、日本仮想化技術が運営するEnterpriseCloud.jpで提供している「OpenStack構築手順書」をオンデマンド書籍として再編集したものです。
- 本書の内容は、執筆時点までの情報を基に執筆されています。紹介したWebサイトやアプリケーション、サービスは変更される可能性があります。
- 本書の内容によって生じる、直接または間接被害について、著者ならびに弊社では、一切の責任を負いかねます。
- 本書中の会社名、製品名、サービス名などは、一般に各社の登録商標、または商標です。なお、本書では©、®、TMは明記していません。

はじめに

　本章は、OpenStack Foundation が公開している公式ドキュメント「OpenStack Installation Guide for Ubuntu 14.04」の内容から、「Block Storage Service」までの構築手順をベースに加筆したものです。OpenStack を Ubuntu Server 14.04.2 ベースで構築する手順を解説しています。Canonical 社が提供する Cloud Archive リポジトリーを使って、OpenStack の最新版 Kilo を導入しましょう。

目次

はじめに iii

第I部　OpenStack 構築編　　　　　　　　　　　　　　　　　　　1

第1章　構築する環境について　　　　　　　　　　　　　　　　3
1.1　環境構築に使用する OS 3
1.2　サーバーの構成について 3
1.3　作成するサーバー（ノード） 4
1.4　ネットワークセグメントの設定 4
1.5　各ノードのネットワーク設定 5
1.6　Ubuntu Server のインストール 6
1.7　Ubuntu Server へのログインと root 権限 8
1.8　設定ファイル等の記述について 8

第2章　OpenStack インストール事前設定　　　　　　　　　　　11
2.1　ネットワークデバイスの設定 11
2.2　ホスト名と静的名前解決の設定 13
2.3　sysctl によるカーネルパラメーターの設定 14
2.4　リポジトリーの設定とパッケージの更新 15
2.5　NTP のインストール 15
2.6　Python 用 MySQL/MariaDB クライアントのインストール 17

目次

第 3 章　sql ノードのインストール前設定　19
3.1　MariaDB のインストール　19

第 4 章　controller ノードのインストール前設定　23
4.1　RabbitMQ のインストール　23
4.2　環境変数設定ファイルの作成　25

第 5 章　Keystone インストールと設定（controller ノード）　27
5.1　データベースの作成・確認　27
5.2　パッケージのインストール　28
5.3　設定の変更　28
5.4　データベースに表を作成　29
5.5　Apache Web サーバーの設定　29
5.6　サービスの再起動と DB の削除　31
5.7　サービスと API エンドポイントの作成　31
5.8　プロジェクトとユーザー、ロールの作成　32
5.9　Keystone の動作の確認　34
5.10　動作の確認　35

第 6 章　Glance のインストールと設定　39
6.1　データベースの作成・確認　39
6.2　ユーザーとサービス、API エンドポイントの作成　40
6.3　パッケージのインストール　41
6.4　設定の変更　41
6.5　データベースにデータ登録　43
6.6　Glance サービスの再起動　43
6.7　動作の確認と使用しないデータベースファイルの削除　43
6.8　イメージの取得と登録　44

第 7 章　Nova のインストールと設定（controller ノード）　47
7.1　データベースの作成・確認　47
7.2　ユーザーとサービス、API エンドポイントの作成　48

7.3	パッケージのインストール	49
7.4	設定変更	49
7.5	データベースにデータを作成	50
7.6	Nova サービスの再起動	51
7.7	使用しないデータベースファイル削除	51
7.8	Glance との通信確認	51

第8章　Nova-Compute のインストール・設定（compute ノード） 53

8.1	パッケージのインストール	53
8.2	設定の変更	53
8.3	Nova-Compute サービスの再起動	54
8.4	controller ノードとの疎通確認	55

第9章　Neutron のインストール・設定（controller ノード） 57

9.1	データベース作成・確認	57
9.2	ユーザーとサービス、API エンドポイントの作成	58
9.3	パッケージのインストール	59
9.4	設定の変更	59
9.5	設定の変更	61
9.6	データベースの作成	62
9.7	ログの確認	62
9.8	使用しないデータベースファイル削除	63
9.9	controller ノードの Neutron と関連サービスの再起動	63
9.10	動作の確認	63

第10章　Neutron のインストール・設定（network ノード） 65

10.1	パッケージのインストール	65
10.2	設定の変更	65
10.3	設定の変更	68
10.4	network ノードの Open vSwitch サービスの再起動	69
10.5	ブリッジデバイス設定	69
10.6	サービスの再起動	69

第 11 章 Neutron のインストール・設定（compute ノード） 73
- 11.1 パッケージのインストール 73
- 11.2 設定の変更 73
- 11.3 compute ノードの Open vSwitch サービスの再起動 75
- 11.4 compute ノードのネットワーク設定 75
- 11.5 compute ノードの Neutron と関連サービスを再起動 75
- 11.6 ログの確認 76
- 11.7 Neutron サービスの動作の確認 76

第 12 章 仮想ネットワーク設定（controller ノード） 77
- 12.1 外部接続ネットワークの設定 77
- 12.2 インスタンス用ネットワークの設定 78
- 12.3 仮想ネットワークルーターの設定 79

第 13 章 仮想ネットワーク設定確認（network ノード） 81
- 13.1 仮想ネットワークルーターの確認 81
- 13.2 仮想ルーターのネームスペースの IP アドレスを確認 81
- 13.3 仮想ゲートウェイの疎通を確認 82
- 13.4 インスタンスの起動確認 82

第 14 章 Cinder のインストール（controller ノード） 85
- 14.1 データベース作成・確認 85
- 14.2 ユーザーとサービス、API エンドポイントの作成 86
- 14.3 パッケージのインストール 88
- 14.4 設定の変更 88
- 14.5 データベースに表を作成 89

（目次冒頭）
- 10.7 ブリッジデバイス設定確認 70
- 10.8 ネットワークインターフェースの設定変更 70
- 10.9 network ノードの再起動 71
- 10.10 ブリッジの設定を確認 71
- 10.11 Neutron サービスの動作確認 71

14.6	Cinder サービスの再起動	89
14.7	使用しないデータベースファイルの削除	89
14.8	イメージ格納用ボリュームの作成	89

第15章　Dashboard インストール・確認（controller ノード）　93

15.1	パッケージのインストール	93
15.2	Dashboard の設定の変更	93
15.3	Dashboard へのアクセス確認	94
15.4	セキュリティグループの設定	94
15.5	キーペアの作成	95
15.6	インスタンスの起動	95
15.7	Floating IP の設定	96
15.8	インスタンスへのアクセス	96

第 II 部　監視環境 構築編　97

第16章　Zabbix のインストール　99

16.1	パッケージのインストール	99
16.2	Zabbix 用データベースの作成	100
16.3	Zabbix サーバーの設定および起動	101
16.4	Zabbix frontend の設定および起動	101

第17章　Hatohol のインストール　105

17.1	インストール	106
17.2	セットアップ	106
17.3	セキュリティ設定の変更	108
17.4	Hatohol による情報の閲覧	109
17.5	Hatohol に Zabbix サーバーを登録	109
17.6	Hatohol で Zabbix サーバーの監視	110
17.7	Hatohol でその他のホストの監視	111
17.8	Hatohol Arm Plugin Interface を使用する場合の操作	113

目次

付録 A　　FAQ フォーラム参加特典について **115**

第Ⅰ部

OpenStack 構築編

第1章 構築する環境について

1.1 環境構築に使用するOS

本書はCanonicalのUbuntu ServerとCloud Archiveリポジトリーのパッケージを使って、OpenStack Kiloを構築する手順を解説したものです。

OSはUbuntu Server 14.04.2 LTS（以下Ubuntu Server）のイメージを使用してインストールします。以下のURLよりイメージをダウンロードし、各サーバーへインストールします。

- http://old-releases.ubuntu.com/releases/14.04.2/ubuntu-14.04.2-server-amd64.iso

このドキュメントはUbuntu Server 14.04.2がリリースされた時期に執筆されたものです。

Ubuntu Server 14.04.2より新しいバージョンを使ってOpenStack Kilo環境を構築する場合は本手順で説明したよりも新しいバージョンのLinux Kernelが提供されるため、「1-6-2 カーネルの更新」の手順は不要です。

1.2 サーバーの構成について

本書はOpenStack環境を4台のサーバー上に構築することを想定しています。

従来の3台構成でのインストールをお望みの場合は、sqlノードでインストールおよび実行している設定、コマンドをcontrollerノードで実行してください。また設定ファイルへのデータベースの追記は、次のようにコントローラーノードを指定すればこの手順に従って構築可能です。

(3 ノード構成時の Keystone データベースの設定記述例)

```
connection = mysql://keystone:password@sql/keystone
↓
connection = mysql://keystone:password@controller/keystone
```

1.3　作成するサーバー（ノード）

今回構築する OpenStack 環境は、以下 4 台のサーバーで構成します。

- sql ノード
  データベースサーバー「MariaDB」の実行用のノードです。データベースはこのノード上で作成します。
- controller ノード
  OpenStack 環境全体を管理するコントローラーとして機能します。
- network ノード
  外部ネットワークとインスタンスの間のネットワークを制御します。
- compute ノード
  仮想マシンインスタンスを実行します。

1.4　ネットワークセグメントの設定

今回は 2 つのネットワークセグメントを用意し構成しています。

- 内部ネットワーク（Instance Tunnels）network ノードと compute ノード間のトンネル用に使用するネットワーク。インターネットへの接続は行えなくても構いません。
- 外部ネットワーク（Management）外部との接続に使用するネットワーク。構築中は apt コマンドを使って外部リポジトリーからパッケージなどをダウンロードするため、インターネット接続が必要となります。

　OpenStack 稼働後は、仮想マシンインスタンスに対し Floating IP アドレスを割り当てることで、外部ネットワークへ接続できます。
　なお、各種 API を外部公開する際にも使用できますが、今回の手順では API の公開は行いません。

IP アドレスは以下の構成で構築されている前提で解説します。

-	外部ネットワーク	内部ネットワーク
インターフェース	eth0	eth1
ネットワーク	10.0.0.0/24	192.168.0.0/24
ゲートウェイ	10.0.0.1	なし
ネームサーバー	10.0.0.1	なし

1.5　各ノードのネットワーク設定

各ノードのネットワーク設定は以下の通りです。

- sql ノード

インターフェース	eth0	eth1
IP アドレス	10.0.0.100	192.168.0.100
ネットマスク	255.255.255.0	255.255.255.0
ゲートウェイ	10.0.0.1	なし
ネームサーバー	10.0.0.1	なし

- controller ノード

インターフェース	eth0	eth1
IP アドレス	10.0.0.101	192.168.0.101
ネットマスク	255.255.255.0	255.255.255.0
ゲートウェイ	10.0.0.1	なし
ネームサーバー	10.0.0.1	なし

- network ノード

インターフェース	eth0	eth1
IP アドレス	10.0.0.102	192.168.0.102
ネットマスク	255.255.255.0	255.255.255.0
ゲートウェイ	10.0.0.1	なし
ネームサーバー	10.0.0.1	なし

- compute ノード

第 1 章　構築する環境について

インターフェース	eth0	eth1
IP アドレス	10.0.0.103	192.168.0.103
ネットマスク	255.255.255.0	255.255.255.0
ゲートウェイ	10.0.0.1	なし
ネームサーバー	10.0.0.1	なし

1.6　Ubuntu Server のインストール

インストール

4 台のサーバーに対し、Ubuntu Server をインストールします。要点は以下の通りです。

- 優先ネットワークインターフェースを eth0 に指定
    - インターネットへ接続するインターフェースは eth0 を使用するため、インストール中は eth0 を優先ネットワークとして指定します。
- パッケージ選択では OpenSSH server のみ選択
    - OS は最小構成でインストールします。
    - compute ノードでは KVM を利用しますが、インストーラでは Virtual machine host のインストールを行わないでください。

【インストール時の設定パラメーター例】

設定項目	設定例
初期起動時の Language	English
起動	Install Ubuntu Server
言語	English - English
地域の設定	other → Asia → Japan
地域の言語	United States - en_US.UTF-8
キーボードレイアウトの認識	No
キーボードの言語	Japanese → Japanese
優先する NIC	eth0: Ethernet
ホスト名	それぞれのノード名 (controller, network, compute1)

1.6 Ubuntu Server のインストール

設定項目	設定例
ユーザ名とパスワード	フルネームで入力
アカウント名	ユーザ名のファーストネームで設定される
パスワード	任意のパスワード
Weak password（出ない場合も）	Yes を選択
ホームの暗号化	任意
タイムゾーン	Asia/Tokyo であることを確認
パーティション設定	Guided - use entire disk and set up LVM
パーティション選択	sda を選択
パーティション書き込み	Yes を選択
パーティションサイズ	デフォルトのまま
変更の書き込み	Yes を選択
HTTP proxy	環境に合わせて任意
アップグレード	No automatic updates を推奨
ソフトウェア	OpenSSH server のみ選択
GRUB	Yes を選択
インストール完了	Continue を選択

筆者注：
Ubuntu インストール時に選択した言語がインストール後も使用されます。
Ubuntu Server の言語で日本語を設定した場合、標準出力や標準エラー出力が文字化けするなど様々な問題が起きますので、言語設定は英語を推奨します。

カーネルの更新

Ubuntu Server 14.04.2 は Linux Kernel 3.16 系のカーネルが利用されます。次のコマンドを実行すると、Ubuntu 15.04 と同等の Linux Kernel 3.19 系のカーネルを Ubuntu 14.04 LTS で利用できます。Linux Kernel 3.19 系ではサーバー向けに様々なパフォーマンスの改善が行われています。必要に応じてアップデートしてください。詳細は 「Ubuntu Wiki の記事 (https://wiki.ubuntu.com/VividVervet/ReleaseNotes/Ja#Linux_kernel_3.19)」をご覧ください。

```
# apt-get install -y linux-headers-generic-lts-vivid linux-image-generic-lts-vivid
```

Linux Kernel 3.16 系のカーネルのアップデートが不要の場合は再起動後に実行します。

```
# apt-get remove -y linux-image-generic-lts-utopic
```

必要に応じて Linux Kernel 3.16 系のカーネルを削除してください。

プロキシーの設定

外部ネットワークとの接続にプロキシーの設定が必要な場合は、apt コマンドを使ってパッ

ケージの照会やダウンロードを行うために次のような設定をする必要があります。

- システムのプロキシー設定

```
# vi /etc/environment
http_proxy="http://proxy.example.com:8080/"
https_proxy="https://proxy.example.com:8080/"
```

- APTのプロキシー設定

```
# vi /etc/apt/apt.conf
Acquire::http::proxy "http://proxy.example.com:8080/";
Acquire::https::proxy "https://proxy.example.com:8080/";
```

より詳細な情報は下記のサイトの情報を確認ください。

- https://help.ubuntu.com/community/AptGet/Howto
- http://gihyo.jp/admin/serial/01/ubuntu-recipe/0331

1.7　Ubuntu Serverへのログインとroot権限

　Ubuntuはデフォルト設定でrootユーザーの利用を許可していないため、root権限が必要となる作業は以下のように行ってください。

- rootユーザーで直接ログインできないので、インストール時に作成したアカウントでログインする
- root権限が必要な場合には、sudoコマンドを使用する
- rootで連続して作業したい場合には、sudo -iコマンドでシェルを起動する

1.8　設定ファイル等の記述について

- 設定ファイルは特別な記述がない限り、必要な設定を抜粋したものです
- 特に変更の必要がない設定項目は省略されています
- [見出し]が付いている場合、その見出しから次の見出しまでの間に設定を記述します
- コメントアウトされていない設定項目が存在する場合には、値を変更してください。多く

1.8 設定ファイル等の記述について

の設定項目は記述が存在しているため、エディタの検索機能で検索することをお勧めします

- 特定のホストでコマンドを実行する場合はコマンドの冒頭にホスト名を記述しています

【設定ファイルの記述例】

```
controller# vi /etc/glance/glance-api.conf ←コマンド冒頭にこのコマンドを実行するホストを記述

[database] ←この見出しから次の見出しまでの間に以下を記述
#connection = sqlite:////var/lib/glance/glance.sqlite    ← 既存設定をコメントアウト
connection = mysql://glance:password@controller/glance   ← 追記


[keystone_authtoken] ← 見出し
#auth_host = 127.0.0.1 ← 既存設定をコメントアウト
auth_host = controller ← 追記

auth_port = 35357
auth_protocol = http
auth_uri = http://controller:5000/v2.0 ← 追記
admin_tenant_name = service ← 変更
admin_user = glance ← 変更
admin_password = password ← 変更
```

第2章 OpenStack インストール事前設定

OpenStack パッケージのインストール前に各々のノードで以下の設定を行います。

- ネットワークデバイスの設定
- ホスト名と静的名前解決の設定
- sysctl によるカーネルパラメーターの設定
- リポジトリーの設定とパッケージの更新
- NTP サーバーのインストール（controller ノードのみ）
- NTP クライアントのインストール
- Python 用 MySQL/MariaDB クライアントのインストール
- MariaDB のインストール（sql ノードのみ）
- RabbitMQ のインストール（controller ノードのみ）

2.1 ネットワークデバイスの設定

各ノードの /etc/network/interfaces を編集し、IP アドレスの設定を行います。

sql ノードの IP アドレスの設定

```
sql# vi /etc/network/interfaces

auto eth0
iface eth0 inet static
    address 10.0.0.100
```

```
        netmask 255.255.255.0
        gateway 10.0.0.1
        dns-nameservers 10.0.0.1

auto eth1
iface eth1 inet static
        address 192.168.0.100
        netmask 255.255.255.0
```

controllerノードのIPアドレスの設定

```
controller# vi /etc/network/interfaces

auto eth0
iface eth0 inet static
        address 10.0.0.101
        netmask 255.255.255.0
        gateway 10.0.0.1
        dns-nameservers 10.0.0.1

auto eth1
iface eth1 inet static
        address 192.168.0.101
        netmask 255.255.255.0
```

networkノードのIPアドレスの設定

```
network# vi /etc/network/interfaces

auto eth0
iface eth0 inet static
        address 10.0.0.102
        netmask 255.255.255.0
        gateway 10.0.0.1
        dns-nameservers 10.0.0.1

auto eth1
iface eth1 inet static
        address 192.168.0.102
        netmask 255.255.255.0
```

computeノードのIPアドレスの設定

```
compute1# vi /etc/network/interfaces

auto eth0
iface eth0 inet static
```

```
        address 10.0.0.103
        netmask 255.255.255.0
        gateway 10.0.0.1
        dns-nameservers 10.0.0.1
auto eth1
 iface eth1 inet static
        address 192.168.0.103
        netmask 255.255.255.0
```

ネットワーク設定反映

各ノードで変更した設定を反映させるため、ホストを再起動します。

```
# shutdown -r now
```

2.2　ホスト名と静的名前解決の設定

各ノードの/etc/hosts に各ノードの IP アドレスとホスト名を記述し、静的名前解決の設定を行います。127.0.1.1 の行はコメントアウトします。

各ノードのホスト名の設定

各ノードのホスト名を hostnamectl コマンドを使って設定します。反映させるためには一度ログインしなおす必要があります。

（例）controller の場合

```
# hostnamectl set-hostname controller
# cat /etc/hostname
controller
```

各ノードの/etc/hosts の設定

すべてのノードで 127.0.1.1 の行をコメントアウトします。またホスト名で名前引きできるように設定します。

（例）controller の場合

```
# vi /etc/hosts
127.0.0.1 localhost
#127.0.1.1 controller  ← 既存設定をコメントアウト
#ext
10.0.0.100 sql
```

```
10.0.0.101 controller
10.0.0.102 network
10.0.0.103 compute
#int
192.168.0.100 sql-int
192.168.0.101 controller-int
192.168.0.102 network-int
192.168.0.103 compute-int
```

2.3　sysctlによるカーネルパラメーターの設定

Linuxのネットワークパケット処理について設定を行います。

networkノードの/etc/sysctl.confの設定

```
network# vi /etc/sysctl.conf
net.ipv4.conf.default.rp_filter=0      ← 1から0に変更
net.ipv4.conf.all.rp_filter=0          ← 1から0に変更
net.ipv4.ip_forward=1
```

sysctlコマンドで設定を適用します。

```
network# sysctl -p
net.ipv4.conf.default.rp_filter = 0
net.ipv4.conf.all.rp_filter = 0
net.ipv4.ip_forward = 1
```

computeノードの/etc/sysctl.confの設定

```
compute# vi /etc/sysctl.conf
net.ipv4.conf.default.rp_filter=0           ← 1から0に変更
net.ipv4.conf.all.rp_filter=0               ← 1から0に変更
net.bridge.bridge-nf-call-iptables=1        ← 追記
net.bridge.bridge-nf-call-ip6tables=1       ← 追記
```

最近のLinux Kernelはbridgeではなくbr_netfilterモジュールを読み込む必要があるので設定します。再起動後もモジュールを読み込んでくれるように/etc/modulesに追記します。詳細は「フォーラムの情報 (http://serverfault.com/questions/697942/centos-6-elrepo-kernel-bridge-issues)」をご覧ください。

```
compute# modprobe br_netfilter
compute# ls /proc/sys/net/bridge
bridge-nf-call-arptables   bridge-nf-filter-pppoe-tagged
```

```
bridge-nf-call-ip6tables    bridge-nf-filter-vlan-tagged
bridge-nf-call-iptables     bridge-nf-pass-vlan-input-dev

compute# echo "br_netfilter" >> /etc/modules
```

sysctl コマンドで設定を適用します。

```
compute# sysctl -p
net.ipv4.conf.default.rp_filter = 0
net.ipv4.conf.all.rp_filter = 0
net.bridge.bridge-nf-call-iptables = 1
net.bridge.bridge-nf-call-ip6tables = 1
```

2.4　リポジトリーの設定とパッケージの更新

各ノードで以下のコマンドを実行し、Kilo 向け Ubuntu Cloud Archive リポジトリーを登録します。

```
# add-apt-repository cloud-archive:kilo
 Ubuntu Cloud Archive for OpenStack Kilo
 More info: https://wiki.ubuntu.com/ServerTeam/CloudArchive
Press [ENTER] to continue or ctrl-c to cancel adding it    ← Enter キーを押す
...
Importing ubuntu-cloud.archive.canonical.com keyring
OK
Processing ubuntu-cloud.archive.canonical.com removal keyring
OK
```

各ノードのシステムをアップデートして再起動します。

```
# apt-get update && apt-get -y dist-upgrade && reboot
```

2.5　NTPのインストール

各ノードで時刻を正確にするために NTP をインストールします。

```
# apt-get install -y ntp
```

controller ノードの/etc/ntp.conf の設定

controller ノードで公開 NTP サーバーと同期する NTP サーバーを構築します。適切な公開 NTP サーバー（ex.ntp.nict.jp etc..）を指定します。ネットワーク内に NTP サーバーがある場

第 2 章 OpenStack インストール事前設定

合はそのサーバーを指定します。

内容変更した場合は設定を適用するため、NTP サービスを再起動します。

```
controller# service ntp restart
```

その他ノードの/etc/ntp.conf の設定

network ノードと compute ノードで controller ノードと同期する NTP サーバーを構築します。

```
network# vi /etc/ntp.conf

#server 0.ubuntu.pool.ntp.org    #デフォルト設定はコメントアウト or 削除
#server 1.ubuntu.pool.ntp.org
#server 2.ubuntu.pool.ntp.org
#server 3.ubuntu.pool.ntp.org
#server ntp.ubuntu.com

server controller iburst
```

設定を適用するため、NTP サービスを再起動します。

```
network# service ntp restart
```

NTP サーバーの動作確認

構築した環境で ntpq -p コマンドを実行して、各 NTP サーバーが同期していることを確認します。

公開 NTP サーバーと同期している controller ノード

```
controller# ntpq -p
     remote           refid      st t when poll reach   delay   offset  jitter
==============================================================================
*ntp-a2.nict.go. .NICT.           1 u    3   64    1   6.569   17.818   0.001
```

controller と同期しているその他ノード

```
compute1# ntpq -p
     remote           refid      st t when poll reach   delay   offset  jitter
==============================================================================
*controller     ntp-a2.nict.go.  2 u  407 1024  377   1.290   -0.329   0.647
```

2.6 Python用MySQL/MariaDBクライアントのインストール

各ノードでPython用のMySQL/MariaDBクライアントをインストールします。

```
# apt-get install -y python-mysqldb
```

Python MySQLライブラリーはMariaDBと互換性があります。

第3章 sqlノードのインストール前設定

3.1 MariaDBのインストール

sqlノードにデータベースサーバーのMariaDBをインストールします。

パッケージのインストール

apt-getコマンドでmariadb-serverパッケージをインストールします。

```
sql# apt-get update
sql# apt-get install -y mariadb-server
```

インストール中にパスワードの入力を要求されますので、MariaDBのrootユーザーに対するパスワードを設定します。本例ではパスワードとして「password」を設定します。

MariaDB設定の変更

MariaDBの設定ファイルmy.cnfを開き以下の設定を変更します。

- バインドアドレスをeth0に割り当てたIPアドレスへ変更
- 文字コードをUTF-8へ変更

別のノードからMariaDBへアクセスできるようにするためバインドアドレスを変更します。加えて使用する文字コードをutf8に変更します。

※文字コードをutf8に変更しないとOpenStackモジュールとデータベース間の通信でエラーが発生します。

第 3 章　sql ノードのインストール前設定

```
sql# vi /etc/mysql/my.cnf

[mysqld]
#bind-address = 127.0.0.1                    ← 既存設定をコメントアウト
bind-address = 10.0.0.100                    ← 追記 (sql ノードの IP アドレス)
default-storage-engine = innodb              ← 追記
innodb_file_per_table                        ← 追記
collation-server = utf8_general_ci           ← 追記
init-connect = 'SET NAMES utf8'              ← 追記
character-set-server = utf8                  ← 追記
```

MariaDB サービスの再起動

変更した設定を反映させるため MariaDB のサービスを再起動します。

```
sql# service mysql restart
```

MariaDB データベースのセキュア化

mysql_secure_installation コマンドを実行すると、データベースのセキュリティを強化できます。必要に応じて設定を行ってください。

- root パスワードの入力

```
sql# mysql_secure_installation
In order to log into MariaDB to secure it, we'll need the current
password for the root user.  If you've just installed MariaDB, and
you haven't set the root password yet, the password will be blank,
so you should just press enter here.
Enter current password for root (enter for none):  password   ← MariaDB の root パス
ワードを入力
```

- root パスワードの変更

```
Setting the root password ensures that nobody can log into the MariaDB
root user without the proper authorisation.
You already have a root password set, so you can safely answer 'n'.
Change the root password? [Y/n]  n
```

- anonymous ユーザーの削除

```
By default, a MariaDB installation has an anonymous user, allowing anyone
to log into MariaDB without having to have a user account created for
them.  This is intended only for testing, and to make the installation
go a bit smoother.  You should remove them before moving into a
production environment.
Remove anonymous users? [Y/n] y
```

- リモートからの root ログインを禁止

本例ではデータベースの操作はすべて sql ノード上で行うことを想定しているため、リモートからの root ログインを禁止します。必要に応じて設定してください。

```
Normally, root should only be allowed to connect from 'localhost'.  This
ensures that someone cannot guess at the root password from the network.
Disallow root login remotely? [Y/n] y
```

- test データベースの削除

```
By default, MariaDB comes with a database named 'test' that anyone can
access.  This is also intended only for testing, and should be removed
before moving into a production environment.
Remove test database and access to it? [Y/n] y
```

- 権限の再読み出し

```
Reloading the privilege tables will ensure that all changes made so far
will take effect immediately.
Reload privilege tables now? [Y/n] y
 ....
Thanks for using MariaDB!
```

MariaDB クライアントのインストール

sql ノード以外のノードに、インストール済みの MariaDB と同様のバージョンの MariaDB クライアントをインストールします。

```
# apt-get update
# apt-get install -y mariadb-client-5.5 mariadb-client-core-5.5
```

mytop のインストール

データベースの状態を確認するため、データベースパフォーマンスモニターツールの mytop

第 3 章 sql ノードのインストール前設定

をインストールします。

```
sql# apt-get update
sql# apt-get install -y mytop
```

利用するには、sql ノードで次のように実行します。ロードアベレージやデータの in/out などの情報を確認できます。

```
sql# mytop --prompt
Password: password      ← MariaDB の root パスワードを入力
```

第4章 controllerノードのインストール前設定

4.1 RabbitMQのインストール

　OpenStackは、オペレーションやステータス情報を各サービス間で連携するためにメッセージブローカーを使用しています。OpenStackではRabbitMQ、Qpid、ZeroMQなど複数のメッセージブローカーサービスに対応しています。本書ではRabbitMQをインストールする例を説明します。

パッケージのインストール

　apt-getコマンドで、rabbitmq-serverパッケージをインストールします。Cloud Archiveリポジトリーのバージョン3.4.3-2は執筆時点のバージョンでは正常に動かないので、標準リポジトリーの最新版をインストールします。

```
controller# apt-get update
controller# apt-cache policy rabbitmq-server
rabbitmq-server:
  Installed: (none)
  Candidate: 3.4.3-2~cloud0
  Version table:
     3.4.3-2~cloud0 0
        500 http://ubuntu-cloud.archive.canonical.com/ubuntu/
trusty-updates/kilo/main amd64 Packages
     3.2.4-1 0
        500 http://jp.archive.ubuntu.com/ubuntu/ trusty/main amd64 Packages
controller# apt-get install -y rabbitmq-server=3.2.4-1   ← 7/2 時点の最新版
controller# apt-mark hold rabbitmq-server                ← バージョンを固定
```

[関連バグ]

- https://bugs.launchpad.net/cloud-archive/+bug/1449392

openstack ユーザーと権限の設定

RabbitMQ にアクセスするユーザーを作成し、パーミッション権限を設定します。

```
# rabbitmqctl add_user openstack password
# rabbitmqctl set_permissions openstack ".*" ".*" ".*"
```

待ち受け IP アドレス・ポートとセキュリティ設定の変更

以下の設定ファイルを作成し、RabbitMQ の待ち受けポートと IP アドレスを定義します。

- 待ち受け設定の追加

```
controller# vi /etc/rabbitmq/rabbitmq-env.conf

RABBITMQ_NODE_IP_ADDRESS=10.0.0.101      ← controller の IP アドレス
RABBITMQ_NODE_PORT=5672
HOSTNAME=controller
```

以下の設定ファイルを作成し、localhost 以外からも RabbitMQ へアクセスできるように設定します。

- リモート認証の許可

```
controller# vi /etc/rabbitmq/rabbitmq.conf

[{rabbit, [{loopback_users, []}]}].
```

RabbitMQ サービス再起動と確認

- ログの確認

メッセージブローカーサービスが正常に動いていないと、OpenStack の各コンポーネントも動作しません。RabbitMQ サービスの再起動と動作確認を行い、確実に動作していることを確認します。

```
controller# service rabbitmq-server restart
controller# tailf /var/log/rabbitmq/rabbit@controller.log
```

※新たなエラーが表示されなければ問題ありません。

- デフォルトユーザー guest で RabbitMQ の Web 管理画面にアクセス

次のように実行して、RabbitMQ の管理画面を有効化します。

```
controller# rabbitmq-plugins enable rabbitmq_management
controller# service rabbitmq-server restart
```

ブラウザーで下記 URL の管理画面にアクセスします。ユーザー:guest パスワード:guest でログインできれば RabbitMQ サーバー自体は正常です。

```
http://controller-node-ipaddress:15672
```

作成した openstack ユーザーでリモートから RabbitMQ の管理画面にログインできないのは正常です。これは openstack ユーザーに administrator 権限が割り振られていないためです。ユーザー権限は「rabbitmqctl list_users」コマンドで確認、任意のユーザーに管理権限を設定するには「rabbitmqctl set_user_tags openstack administrator」のように実行するとログイン可能になります。

4.2　環境変数設定ファイルの作成

admin 環境変数設定ファイル作成

admin ユーザー用環境変数設定ファイルを作成します。

```
controller# vi ~/admin-openrc.sh

export OS_PROJECT_DOMAIN_ID=default
export OS_USER_DOMAIN_ID=default
export OS_PROJECT_NAME=admin
export OS_TENANT_NAME=admin
export OS_USERNAME=admin
export OS_PASSWORD=password
export OS_AUTH_URL=http://controller:35357/v3
export PS1='\u@\h \W(admin)\$ '
```

demo 環境変数設定ファイル作成

demo ユーザー用環境変数設定ファイルを作成します。

```
controller# vi ~/demo-openrc.sh

export OS_PROJECT_DOMAIN_ID=default
export OS_USER_DOMAIN_ID=default
```

第 4 章 controller ノードのインストール前設定

```
export OS_PROJECT_NAME=demo
export OS_TENANT_NAME=demo
export OS_USERNAME=demo
export OS_PASSWORD=password
export OS_AUTH_URL=http://controller:5000/v3
export PS1='\u@\h \W(demo)\$ '
```

第5章 Keystoneインストールと設定（controllerノード）

各サービス間の連携時に使用する認証IDサービスKeystoneのインストールと設定を行います。

5.1 データベースの作成・確認

Keystoneで使用するデータベースを作成します。

データベースの作成

MariaDBにデータベースkeystoneを作成します。

```
sql# mysql -u root -p << EOF
CREATE DATABASE keystone;
GRANT ALL PRIVILEGES ON keystone.* TO 'keystone'@'localhost' \
IDENTIFIED BY 'password';
GRANT ALL PRIVILEGES ON keystone.* TO 'keystone'@'%' \
IDENTIFIED BY 'password';
EOF
Enter password: ← MariaDBのrootパスワードpasswordを入力
```

データベースの確認

sqlノードにユーザーkeystoneでログインしデータベースの閲覧が可能であることを確認します。

第 5 章　Keystone インストールと設定（controller ノード）

```
controller# mysql -h sql -u keystone -p
Enter password:    ← MariaDB の keystone パスワード password を入力
...
Type 'help;' or '\h' for help. Type '\c' to clear the current input statement.

MariaDB [(none)]> show databases;
+--------------------+
| Database           |
+--------------------+
| information_schema |
| keystone           |
+--------------------+
2 rows in set (0.00 sec)
```

admin_token の決定

　Keystone の admin_token に設定するトークン文字列を次のようなコマンドを実行して決定します。出力される結果はランダムな英数字になります。

```
controller# openssl rand -hex 10
64de11ce5f875e977081
```

5.2　パッケージのインストール

　Keystone のインストール時にサービスの自動起動が行われないようにするため、以下のように実行します。

```
controller# echo "manual" > /etc/init/keystone.override
```

　apt-get コマンドで keystone パッケージをインストールします。

```
controller# apt-get update
controller# apt-get install -y keystone python-openstackclient apache2
libapache2-mod-wsgi memcached python-memcache
```

5.3　設定の変更

　keystone の設定ファイルを変更します。

```
controller# vi /etc/keystone/keystone.conf

[DEFAULT]
admin_token = 64de11ce5f875e977081      ← 追記 (5-1-3 で出力されたキーを入力)
```

```
log_dir = /var/log/keystone           ← 設定されていることを確認
verbose = True              ← 追記 (詳細なログを出力する)
...
[database]
#connection = sqlite:////var/lib/keystone/keystone.db    ← 既存設定をコメントアウト
connection = mysql://keystone:password@sql/keystone      ← 追記
...
[memcache]
...
servers = localhost:11211                                ← アンコメント
...
[token]
provider = keystone.token.providers.uuid.Provider        ← アンコメント
driver = keystone.token.persistence.backends.memcache.Token  ← 追記
...
[revoke]
...
driver = keystone.contrib.revoke.backends.sql.Revoke     ← アンコメント
```

次のコマンドを実行して正しく設定を行ったか確認します。

```
controller# less /etc/keystone/keystone.conf | grep -v "^\s*$" | grep -v "^\s*#"
```

5.4 データベースに表を作成

```
controller# su -s /bin/sh -c "keystone-manage db_sync" keystone
```

5.5 Apache Webサーバーの設定

- controllerノードの/etc/apache2/apache2.confのServerNameにcontrollerノードのホスト名を設定します。

```
ServerName controller
```

- controllerノードの/etc/apache2/sites-available/wsgi-keystone.confを作成して、次の内容を記述します。

```
Listen 5000
Listen 35357
<VirtualHost *:5000>
  WSGIDaemonProcess keystone-public processes=5 threads=1 user=keystone
```

第 5 章　Keystone インストールと設定（controller ノード）

```
display-name=%{GROUP}
  WSGIProcessGroup keystone-public
  WSGIScriptAlias / /var/www/cgi-bin/keystone/main
  WSGIApplicationGroup %{GLOBAL}
  WSGIPassAuthorization On
    <IfVersion >= 2.4>
      ErrorLogFormat "%{cu}t %M"
    </IfVersion>
  LogLevel info
  ErrorLog /var/log/apache2/keystone-error.log
  CustomLog /var/log/apache2/keystone-access.log combined
</VirtualHost>
<VirtualHost *:35357>
  WSGIDaemonProcess keystone-admin processes=5 threads=1 user=keystone
display-name=%{GROUP}
  WSGIProcessGroup keystone-admin
  WSGIScriptAlias / /var/www/cgi-bin/keystone/admin
  WSGIApplicationGroup %{GLOBAL}
  WSGIPassAuthorization On
    <IfVersion >= 2.4>
      ErrorLogFormat "%{cu}t %M"
    </IfVersion>
  LogLevel info
  ErrorLog /var/log/apache2/keystone-error.log
  CustomLog /var/log/apache2/keystone-access.log combined
</VirtualHost>
```

- バーチャルホストで Identity service を有効に設定します。

```
controller# ln -s /etc/apache2/sites-available/wsgi-keystone.conf /etc/apache2/sites-enabled
```

- WSGI コンポーネント用のディレクトリーを作成します。

```
controller# mkdir -p /var/www/cgi-bin/keystone
```

- WSGI コンポーネントを Upstream リポジトリーからコピーし、先ほどのディレクトリーに展開します。

```
controller# curl http://git.openstack.org/cgit/openstack/keystone/plain/httpd/keystone.py?h=stable/kilo | tee /var/www/cgi-bin/keystone/main /var/www/cgi-bin/keystone/admin
```

- ディレクトリーとファイルのパーミッションを修正します。

```
controller# chown -R keystone:keystone /var/www/cgi-bin/keystone
controller# chmod 755 /var/www/cgi-bin/keystone/*
```

5.6 サービスの再起動とDBの削除

- Apache Web サーバーを再起動します。

```
controller# service apache2 restart
```

- パッケージのインストール時に作成される不要なSQLiteファイルを削除します。

```
controller# rm /var/lib/keystone/keystone.db
```

5.7 サービスとAPIエンドポイントの作成

以下コマンドでサービスと API エンドポイントを設定します。

- 環境変数の設定

```
controller# export OS_TOKEN=64de11ce5f875e977081    ← 追記 (5-1-3で出力されたキーを入力)
controller# export OS_URL=http://controller:35357/v2.0
```

- サービスを作成

```
controller# openstack service create \
  --name keystone --description "OpenStack Identity" identity
+-------------+----------------------------------+
| Field       | Value                            |
+-------------+----------------------------------+
| description | OpenStack Identity               |
| enabled     | True                             |
| id          | 492157c4ba4c432995a6ebbf579b8654 |
| name        | keystone                         |
| type        | identity                         |
+-------------+----------------------------------+
```

- APIエンドポイントを作成

第 5 章　Keystone インストールと設定（controller ノード）

```
controller# openstack endpoint create \
--publicurl http://controller:5000/v2.0 \
--internalurl http://controller:5000/v2.0 \
--adminurl http://controller:35357/v2.0 \
--region RegionOne identity
+--------------+----------------------------------+
| Field        | Value                            |
+--------------+----------------------------------+
| adminurl     | http://controller:35357/v2.0     |
| id           | e10594bbf242482c86e8a9076c41957e |
| internalurl  | http://controller:5000/v2.0      |
| publicurl    | http://controller:5000/v2.0      |
| region       | RegionOne                        |
| service_id   | 492157c4ba4c432995a6ebbf579b8654 |
| service_name | keystone                         |
| service_type | identity                         |
+--------------+----------------------------------+
```

5.8　プロジェクトとユーザー、ロールの作成

以下コマンドで認証情報（テナント・ユーザー・ロール）を設定します。

- admin プロジェクトの作成

```
controller# openstack project create --description "Admin Project" admin
+-------------+----------------------------------+
| Field       | Value                            |
+-------------+----------------------------------+
| description | Admin Project                    |
| enabled     | True                             |
| id          | 218010a87fe5477bba7f5e25c8211614 |
| name        | admin                            |
+-------------+----------------------------------+
```

- admin ユーザーの作成

```
controller# openstack user create --password-prompt admin
User Password: password   #admin ユーザーのパスワードを設定 (本例は password を設定)
Repeat User Password: password
+----------+----------------------------------+
| Field    | Value                            |
+----------+----------------------------------+
| email    | None                             |
| enabled  | True                             |
| id       | 9caffb5dc1d749c5b3e9493139fe8598 |
| name     | admin                            |
```

```
| username | admin                             |
+----------+-----------------------------------+
```

- admin ロールの作成

```
controller# openstack role create admin
+-------+----------------------------------+
| Field | Value                            |
+-------+----------------------------------+
| id    | 9212e4ba1d07418a97fb4eaaaa275334 |
| name  | admin                            |
+-------+----------------------------------+
```

- admin プロジェクトとユーザーに admin ロールを追加

```
controller# openstack role add --project admin --user admin admin
+-------+----------------------------------+
| Field | Value                            |
+-------+----------------------------------+
| id    | 9212e4ba1d07418a97fb4eaaaa275334 |
| name  | admin                            |
+-------+----------------------------------+
```

- service プロジェクトを作成

```
controller# openstack project create --description "Service Project" service
+-------------+----------------------------------+
| Field       | Value                            |
+-------------+----------------------------------+
| description | Service Project                  |
| enabled     | True                             |
| id          | 5b786e6b78d248df91b8722f513e38d2 |
| name        | service                          |
+-------------+----------------------------------+
```

- demo プロジェクトの作成

```
controller# openstack project create --description "Demo Project" demo
+-------------+----------------------------------+
| Field       | Value                            |
+-------------+----------------------------------+
| description | Demo Project                     |
| enabled     | True                             |
| id          | 3ed2437abf474a37b305338666d9fafa |
| name        | demo                             |
```

第 5 章 Keystone インストールと設定（controller ノード）

```
+------------+------------------------------+
```

- demo ユーザーの作成

```
controller# openstack user create --password-prompt demo
User Password: password    #demo ユーザーのパスワードを設定 (本例は password を設定)
Repeat User Password: password
+----------+----------------------------------+
| Field    | Value                            |
+----------+----------------------------------+
| email    | None                             |
| enabled  | True                             |
| id       | 81b6c592d5a847a1b0ee8740d14a2e3b |
| name     | demo                             |
| username | demo                             |
+----------+----------------------------------+
```

- user ロールの作成

```
controller# openstack role create user
+-------+----------------------------------+
| Field | Value                            |
+-------+----------------------------------+
| id    | da8e8598734a47bd9da2404dad7b4884 |
| name  | user                             |
+-------+----------------------------------+
```

- demo プロジェクトと demo ユーザーに user ロールを追加

```
controller# openstack role add --project demo --user demo user
+-------+----------------------------------+
| Field | Value                            |
+-------+----------------------------------+
| id    | da8e8598734a47bd9da2404dad7b4884 |
| name  | user                             |
+-------+----------------------------------+
```

5.9　Keystoneの動作の確認

　他のサービスをインストールする前に Identity サービスが正しく構築、設定されたか動作を検証します。

- セキュリティを確保するため、一時認証トークンメカニズムを無効化します。

- /etc/keystone/keystone-paste.ini を開き、[pipeline:public_api] と [pipeline:admin_api] と [pipeline:api_v3] セクションの pipeline 行から、admin_token_auth を取り除きます。

```
[pipeline:public_api]
pipeline = sizelimit url_normalize request_id build_auth_context token_auth
json_body ec2_extension user_crud_extension public_service
...
[pipeline:admin_api]
pipeline = sizelimit url_normalize request_id build_auth_context token_auth
json_body ec2_extension s3_extension crud_extension admin_service
...
[pipeline:api_v3]
pipeline = sizelimit url_normalize request_id build_auth_context token_auth
json_body ec2_extension_v3 s3_extension simple_cert_extension revoke_extension
federation_extension oauth1_extension endpoint_filter_extension
endpoint_policy_extension service_v3
```

- Keystone への作成が完了したら環境変数をリセットします。

```
controller# unset OS_TOKEN OS_URL
```

5.10　動作の確認

　動作確認のため admin および demo テナントに対し認証トークンを要求してみます。admin、demo ユーザーのパスワードを入力する必要があります。

- admin ユーザーとして、Identity バージョン 2.0 API から管理トークンを要求します。

```
controller# openstack --os-auth-url http://controller:35357 \
  --os-project-name admin --os-username admin --os-auth-type password \
  token issue
Password:
+------------+----------------------------------+
| Field      | Value                            |
+------------+----------------------------------+
| expires    | 2015-06-22T10:02:36Z             |
| id         | 12ca032c6b914a7382e93d9b371b52d8 |
| project_id | 218010a87fe5477bba7f5e25c8211614 |
| user_id    | 9caffb5dc1d749c5b3e9493139fe8598 |
+------------+----------------------------------+
```

　正常に応答が返ってくると、/var/log/apache2/keystone-access.log に HTTP 200 と記録さ

第 5 章　Keystone インストールと設定（controller ノード）

れます。正常に応答がない場合は/var/log/apache2/keystone-error.log を確認しましょう。

```
...
10.0.0.101 - - [23/Jun/2015:09:55:21 +0900] "GET / HTTP/1.1" 300 789 "-"
"python-keystoneclient"
10.0.0.101 - - [23/Jun/2015:09:55:24 +0900] "POST /v2.0/tokens HTTP/1.1" 200 1070
"-" "python-keystoneclient"
```

- admin ユーザーとして、Identity バージョン 3.0 API から管理トークンを要求します。

```
controller# openstack --os-auth-url http://controller:35357 \
  --os-project-domain-id default --os-user-domain-id default \
  --os-project-name admin --os-username admin --os-auth-type password \
  token issue
Password:
+------------+----------------------------------+
| Field      | Value                            |
+------------+----------------------------------+
| expires    | 2015-06-22T10:05:59.140858Z      |
| id         | 5bc15107d6604841b2b006d30c5b94bb |
| project_id | 218010a87fe5477bba7f5e25c8211614 |
| user_id    | 9caffb5dc1d749c5b3e9493139fe8598 |
+------------+----------------------------------+
```

- admin ユーザーで管理ユーザー専用のコマンドを使って、作成したプロジェクトを表示できることを確認します。

```
controller# openstack --os-auth-url http://controller:35357 \
  --os-project-name admin --os-username admin --os-auth-type password \
  project list
Password:
+----------------------------------+---------+
| ID                               | Name    |
+----------------------------------+---------+
| 218010a87fe5477bba7f5e25c8211614 | admin   |
| 3ed2437abf474a37b305338666d9fafa | demo    |
| 5b786e6b78d248df91b8722f513e38d2 | service |
+----------------------------------+---------+
```

- admin ユーザーでユーザーを一覧表示して、先に作成したユーザーが含まれることを確認します。

```
controller# openstack --os-auth-url http://controller:35357 \
  --os-project-name admin --os-username admin --os-auth-type password \
  user list
Password:
+----------------------------------+-------+
```

```
| ID                               | Name  |
+----------------------------------+-------+
| 9caffb5dc1d749c5b3e9493139fe8598 | admin |
| 81b6c592d5a847a1b0ee8740d14a2e3b | demo  |
+----------------------------------+-------+
```

- admin ユーザーでロールを一覧表示して、先に作成したロールが含まれることを確認します。

```
controller# openstack --os-auth-url http://controller:35357 \
  --os-project-name admin --os-username admin --os-auth-type password \
  role list
Password:
+----------------------------------+-------+
| ID                               | Name  |
+----------------------------------+-------+
| 9212e4ba1d07418a97fb4eaaaa275334 | admin |
| da8e8598734a47bd9da2404dad7b4884 | user  |
+----------------------------------+-------+
```

- demo ユーザーとして、Identity バージョン 3 API から管理トークンを要求します。

```
controller# openstack --os-auth-url http://controller:5000 \
  --os-project-domain-id default --os-user-domain-id default \
  --os-project-name demo --os-username demo --os-auth-type password \
  token issue
Password:
+------------+----------------------------------+
| Field      | Value                            |
+------------+----------------------------------+
| expires    | 2015-06-22T10:17:22.735793Z      |
| id         | 658a14e8a2bd49f3aea4ab8208ca1ae5 |
| project_id | 3ed2437abf474a37b305338666d9fafa |
| user_id    | 81b6c592d5a847a1b0ee8740d14a2e3b |
+------------+----------------------------------+
```

- demo ユーザーでは管理権限が必要なコマンド、例えばユーザー一覧の表示を実施するとエラーになることを確認します。

```
controller# openstack --os-auth-url http://controller:5000 \
  --os-project-domain-id default --os-user-domain-id default \
  --os-project-name demo --os-username demo --os-auth-type password \
  user list
Password:
ERROR: openstack You are not authorized to perform the requested action:
admin_required (HTTP 403)
```

第6章 Glance のインストールと設定

6.1 データベースの作成・確認

データベース作成

MariaDB にデータベース glance を作成します。

```
sql# mysql -u root -p << EOF
CREATE DATABASE glance;
GRANT ALL PRIVILEGES ON glance.* TO 'glance'@'localhost' \
 IDENTIFIED BY 'password';
GRANT ALL PRIVILEGES ON glance.* TO 'glance'@'%' \
IDENTIFIED BY 'password';
EOF
Enter password: ← MariaDB の root パスワード password を入力
```

データベースの確認

ユーザー glance でログインしデータベースの閲覧が可能であることを確認します。

```
controller# mysql -h sql -u glance -p
Enter password: ← MariaDB の glance パスワード password を入力
...
Type 'help;' or '\h' for help. Type '\c' to clear the current input statement.

MariaDB [(none)]> show databases;
+--------------------+
| Database           |
+--------------------+
| information_schema |
| glance             |
```

```
+-------------------+
2 rows in set (0.00 sec)
```

6.2 ユーザーとサービス、APIエンドポイントの作成

以下コマンドで認証情報を読み込んだあと、サービスとAPIエンドポイントを設定します。

- 環境変数ファイルの読み込み

admin-openrc.sh を読み込むと次のように出力が変化します。

```
controller# source admin-openrc.sh
controller ~(admin)#
```

- glance ユーザーの作成

```
controller# openstack user create --password-prompt glance
User Password: password   #glance ユーザーのパスワードを設定 (本例は password を設定)
Repeat User Password: password
+----------+----------------------------------+
| Field    | Value                            |
+----------+----------------------------------+
| email    | None                             |
| enabled  | True                             |
| id       | 7cd03c39bf584d49902371154b71c6fc |
| name     | glance                           |
| username | glance                           |
+----------+----------------------------------+
```

- admin ロールを glance ユーザーと service プロジェクトに追加

```
controller# openstack role add --project service --user glance admin
+-------+----------------------------------+
| Field | Value                            |
+-------+----------------------------------+
| id    | 9212e4ba1d07418a97fb4eaaaa275334 |
| name  | admin                            |
+-------+----------------------------------+
```

- サービスの作成

```
controller# openstack service create --name glance \
--description "OpenStack Image service" image
+-------------+----------------------------------+
| Field       | Value                            |
+-------------+----------------------------------+
| description | OpenStack Image service          |
| enabled     | True                             |
| id          | dbef33496c2e4c8aa7127077677242b9 |
| name        | glance                           |
| type        | image                            |
+-------------+----------------------------------+
```

- サービスエンドポイントの作成

```
controller# openstack endpoint create \
--publicurl http://controller:9292 \
--internalurl http://controller:9292 \
--adminurl http://controller:9292 \
--region RegionOne image
+--------------+----------------------------------+
| Field        | Value                            |
+--------------+----------------------------------+
| adminurl     | http://controller:9292           |
| id           | 95206098f81d42b7b50a42e44d4f5acd |
| internalurl  | http://controller:9292           |
| publicurl    | http://controller:9292           |
| region       | RegionOne                        |
| service_id   | dbef33496c2e4c8aa7127077677242b9 |
| service_name | glance                           |
| service_type | image                            |
+--------------+----------------------------------+
```

6.3　パッケージのインストール

apt-get コマンドで glance と glance クライアントパッケージをインストールします。

```
controller# apt-get update
controller# apt-get install -y glance python-glanceclient
```

6.4　設定の変更

　Glance の設定を行います。glance-api.conf、glance-registry.conf ともに、[keystone_authtoken] に追記した設定以外のパラメーターはコメントアウトします。

第 6 章　Glance のインストールと設定

```
controller# vi /etc/glance/glance-api.conf

[DEFAULT]
...
verbose = True                        ← 追記
...
notification_driver = noop            ← アンコメント
rpc_backend = 'rabbit'                ← アンコメント
rabbit_host = controller              ← 変更
rabbit_userid = openstack             ← 変更
rabbit_password = password            ← 変更
...

[database]
#sqlite_db = /var/lib/glance/glance.sqlite        ← 既存設定をコメントアウト
connection = mysql://glance:password@sql/glance   ← 追記

[keystone_authtoken]（既存の設定はコメントアウトし、以下を追記）
...
revocation_cache_time = 10
auth_uri = http://controller:5000
auth_url = http://controller:35357
auth_plugin = password
project_domain_id = default
user_domain_id = default
project_name = service
username = glance
password = password        ← glance ユーザーのパスワード (5-2 で設定したもの)

[paste_deploy]
flavor = keystone                     ← 追記

[glance_store]
default_store =file                                     ← 設定されていることを確認
filesystem_store_datadir = /var/lib/glance/images/      ← 設定されていることを確認
```

次のコマンドを実行して正しく設定を行ったか確認します。

```
controller# less /etc/glance/glance-api.conf | grep -v "^\s*$" | grep -v "^\s*#"
```

```
controller# vi /etc/glance/glance-registry.conf

[DEFAULT]
...
verbose = True                        ← 追記
...
notification_driver = noop            ← アンコメント
rpc_backend = 'rabbit'                ← アンコメント
...
rabbit_host = controller              ← 変更
rabbit_userid = openstack             ← 変更
rabbit_password = password            ← 変更
```

```
...
[database]
#sqlite_db = /var/lib/glance/glance.sqlite          ← 既存設定をコメントアウト
connection = mysql://glance:password@sql/glance     ← 追記

[keystone_authtoken]（既存の設定はコメントアウトし、以下を追記）
...
auth_uri = http://controller:5000
auth_url = http://controller:35357
auth_plugin = password
project_domain_id = default
user_domain_id = default
project_name = service
username = glance
password = password          ← glance ユーザーのパスワード (5-2 で設定したもの)

[paste_deploy]
flavor = keystone            ← 追記
```

次のコマンドを実行して正しく設定を行ったか確認します。

```
controller# less /etc/glance/glance-registry.conf | grep -v "^\s*$" | grep -v "^\s*#"
```

6.5　データベースにデータ登録

下記コマンドにて glance データベースのセットアップを行います。

```
controller# su -s /bin/sh -c "glance-manage db_sync" glance
```

6.6　Glance サービスの再起動

設定を反映させるため、Glance サービスを再起動します。

```
controller# service glance-registry restart && service glance-api restart
```

6.7　動作の確認と使用しないデータベースファイルの削除

サービスの再起動後、ログを参照し Glance Registry と Glance API サービスでエラーが起きていないことを確認します。

第 6 章　Glance のインストールと設定

```
controller# tailf /var/log/glance/glance-api.log
controller# tailf /var/log/glance/glance-registry.log
```

インストール直後は作られていない場合が多いですが、コマンドを実行して glance.sqlite を削除します。

```
controller# rm /var/lib/glance/glance.sqlite
```

6.8　イメージの取得と登録

Glance へインスタンス用仮想マシンイメージを登録します。ここでは、クラウド環境で主にテスト用途で利用される Linux ディストリビューション CirrOS を登録します。

環境変数の設定

Image service に API バージョン 2.0 でアクセスするため、スクリプトを修正して読み込み直します。

```
controller# cd
controller# echo "export OS_IMAGE_API_VERSION=2" | tee -a admin-openrc.sh demo-openrc.sh
controller# source admin-openrc.sh
```

イメージ取得

CirrOS の Web サイトより仮想マシンイメージをダウンロードします。

```
controller# wget http://download.cirros-cloud.net/0.3.4/cirros-0.3.4-x86_64-disk.img
```

イメージ登録

ダウンロードした仮想マシンイメージを Glance に登録します。

```
controller# glance image-create --name "cirros-0.3.4-x86_64" --file
cirros-0.3.4-x86_64-disk.img --disk-format qcow2 --container-format bare \
 --visibility public
+------------------+--------------------------------------+
| Property         | Value                                |
+------------------+--------------------------------------+
| checksum         | ee1eca47dc88f4879d8a229cc70a07c6     |
| container_format | bare                                 |
| created_at       | 2015-06-23T02:50:54Z                 |
```

```
| disk_format    | qcow2                                |
| id             | 390d2978-4a97-4d27-be6e-32642f7a3789 |
| min_disk       | 0                                    |
| min_ram        | 0                                    |
| name           | cirros-0.3.4-x86_64                  |
| owner          | 218010a87fe5477bba7f5e25c8211614     |
| protected      | False                                |
| size           | 13287936                             |
| status         | active                               |
| tags           | []                                   |
| updated_at     | 2015-06-23T02:50:54Z                 |
| virtual_size   | None                                 |
| visibility     | public                               |
+----------------+--------------------------------------+
```

イメージ登録確認

仮想マシンイメージが正しく登録されたか確認します。

```
controller# glance image-list
+--------------------------------------+---------------------+
| ID                                   | Name                |
+--------------------------------------+---------------------+
| 390d2978-4a97-4d27-be6e-32642f7a3789 | cirros-0.3.4-x86_64 |
+--------------------------------------+---------------------+
```

第7章 Novaのインストールと設定(controllerノード)

7.1 データベースの作成・確認

データベースの作成

MariaDBにデータベースnovaを作成します。

```
sql# mysql -u root -p << EOF
CREATE DATABASE nova;
GRANT ALL PRIVILEGES ON nova.* TO 'nova'@'localhost' \
IDENTIFIED BY 'password';
GRANT ALL PRIVILEGES ON nova.* TO 'nova'@'%' \
IDENTIFIED BY 'password';
EOF
Enter password:            ← MariaDBのrootパスワードpasswordを入力
```

データベースの作成確認

※ユーザーnovaでログインしデータベースの閲覧が可能であることを確認します。

```
controller# mysql -h sql -u nova -p
Enter password: ← MariaDBのnovaパスワードpasswordを入力
...
Type 'help;' or '\h' for help. Type '\c' to clear the current input statement.

MariaDB [(none)]> show databases;
+--------------------+
| Database           |
+--------------------+
| information_schema |
| nova               |
```

```
+-------------------+
2 rows in set (0.00 sec)
```

7.2　ユーザーとサービス、APIエンドポイントの作成

以下コマンドで認証情報を読み込んだあと、サービスとAPIエンドポイントを設定します。

- 環境変数ファイルの読み込み

```
controller# source admin-openrc.sh
```

- nova ユーザーの作成

```
controller# openstack user create --password-prompt nova
User Password: password   #nova ユーザーのパスワードを設定 (本例は password を設定)
Repeat User Password: password
+----------+----------------------------------+
| Field    | Value                            |
+----------+----------------------------------+
| email    | None                             |
| enabled  | True                             |
| id       | 186f2d77f0664d7b81b304ea6cb24660 |
| name     | nova                             |
| username | nova                             |
+----------+----------------------------------+
```

- nova ユーザーを admin ロールに追加

```
controller# openstack role add --project service --user nova admin
+-------+----------------------------------+
| Field | Value                            |
+-------+----------------------------------+
| id    | 9212e4ba1d07418a97fb4eaaaa275334 |
| name  | admin                            |
+-------+----------------------------------+
```

- nova サービスの作成

```
controller# openstack service create --name nova --description "OpenStack Compute"
compute
+-------------+----------------------------------+
| Field       | Value                            |
+-------------+----------------------------------+
| description | OpenStack Compute                |
| enabled     | True                             |
| id          | a0f7280ea95a4c1298764c9e12f99b49 |
| name        | nova                             |
| type        | compute                          |
+-------------+----------------------------------+
```

- Compute サービスの API エンドポイントを作成

```
controller# openstack endpoint create \
--publicurl http://controller:8774/v2/%\(tenant_id\)s \
--internalurl http://controller:8774/v2/%\(tenant_id\)s \
--adminurl http://controller:8774/v2/%\(tenant_id\)s \
--region RegionOne compute
+--------------+---------------------------------------+
| Field        | Value                                 |
+--------------+---------------------------------------+
| adminurl     | http://controller:8774/v2/%(tenant_id)s |
| id           | 152d36c70a50474ca64deb4c2221aa5f       |
| internalurl  | http://controller:8774/v2/%(tenant_id)s |
| publicurl    | http://controller:8774/v2/%(tenant_id)s |
| region       | RegionOne                             |
| service_id   | a0f7280ea95a4c1298764c9e12f99b49      |
| service_name | nova                                  |
| service_type | compute                               |
+--------------+---------------------------------------+
```

7.3 パッケージのインストール

apt-get コマンドで Nova 関連のパッケージをインストールします。

```
controller# apt-get update
controller# apt-get install -y nova-api nova-cert nova-conductor nova-consoleauth nova-novncproxy \
nova-scheduler python-novaclient
```

7.4 設定変更

nova.conf に下記の設定を追記します。

```
controller# vi /etc/nova/nova.conf

[DEFAULT]
...
rpc_backend = rabbit              ←追記
auth_strategy = keystone          ←追記

# controller ノードの IP アドレス:10.0.0.101
my_ip = 10.0.0.101                                 ←追記
vncserver_listen = 10.0.0.101                      ←追記
vncserver_proxyclient_address = 10.0.0.101         ←追記

(↓これ以下追記↓)
[database]
connection = mysql://nova:password@sql/nova

[oslo_messaging_rabbit]
rabbit_host = controller
rabbit_userid = openstack
rabbit_password = password

[keystone_authtoken]
auth_uri = http://controller:5000
auth_url = http://controller:35357
auth_plugin = password
project_domain_id = default
user_domain_id = default
project_name = service
username = nova
password = password       ← nova ユーザーのパスワード (6-2 で設定したもの)

[glance]
host = controller

[oslo_concurrency]
lock_path = /var/lib/nova/tmp
```

次のコマンドを実行して正しく設定を行ったか確認します。

```
controller# less /etc/nova/nova.conf | grep -v "^\s*$" | grep -v "^\s*#"
```

7.5　データベースにデータを作成

下記コマンドにて nova データベースのセットアップを行います。

```
controller# su -s /bin/sh -c "nova-manage db sync" nova
```

7.6　Novaサービスの再起動

設定を反映させるため、Nova のサービスを再起動します。

```
controller# service nova-api restart && service nova-cert restart && \
service nova-consoleauth restart && service nova-scheduler restart && \
service nova-conductor restart && service nova-novncproxy restart
```

7.7　使用しないデータベースファイル削除

データベースは MariaDB を使用するため、使用しない SQLite ファイルを削除します。

```
controller# rm /var/lib/nova/nova.sqlite
```

7.8　Glanceとの通信確認

Nova のコマンドラインインターフェースで Glance と相互に通信できているかを確認します。

```
controller# nova image-list
+--------------------------------------+---------------------+--------+--------+
| ID                                   | Name                | Status | Server |
+--------------------------------------+---------------------+--------+--------+
| 390d2978-4a97-4d27-be6e-32642f7a3789 | cirros-0.3.4-x86_64 | ACTIVE |        |
+--------------------------------------+---------------------+--------+--------+
```

※ Glance に登録した CirrOS イメージが表示できていれば問題ありません。

第8章 Nova-Compute のインストール・設定 (computeノード)

8.1 パッケージのインストール

```
compute# apt-get update
compute# apt-get install -y nova-compute sysfsutils
```

8.2 設定の変更

nova の設定ファイルを変更します。

```
compute# vi /etc/nova/nova.conf

[DEFAULT]
...
rpc_backend = rabbit
auth_strategy = keystone

my_ip = 10.0.0.103              ← IPアドレスで指定
vnc_enabled = True
vncserver_listen = 0.0.0.0
vncserver_proxyclient_address = 10.0.0.103   ← IPアドレスで指定
novncproxy_base_url = http://controller:6080/vnc_auto.html
vnc_keymap = ja                              ← 日本語キーボードの設定

[oslo_messaging_rabbit]
rabbit_host = controller
rabbit_userid = openstack
rabbit_password = password
```

第 8 章　Nova-Compute のインストール・設定（compute ノード）

```
[keystone_authtoken]
auth_uri = http://controller:5000
auth_url = http://controller:35357
auth_plugin = password
project_domain_id = default
user_domain_id = default
project_name = service
username = nova
password = password       ← nova ユーザーのパスワード (6-2 で設定したもの)

[glance]
host = controller

[oslo_concurrency]
lock_path = /var/lib/nova/tmp
```

次のコマンドを実行して正しく設定を行ったか確認します。

```
compute# less /etc/nova/nova.conf | grep -v "^\s*$" | grep -v "^\s*#"
```

　nova-compute の設定ファイルを開き、KVM を利用するように設定変更します。「egrep -c '(vmx|svm)' /proc/cpuinfo」とコマンドを実行して、0 と出たら qemu、0 以上の数字が出たら kvm を virt_type パラメーターに設定する必要があります。

　まず次のようにコマンドを実行し、KVM が動く環境であることを確認します。CPU が VMX もしくは SVM 対応であるか、コア数がいくつかを出力しています。0 と表示される場合は後述の設定で virt_type = qemu を設定します。

```
# cat /proc/cpuinfo |egrep 'vmx|svm'|wc -l
4
```

　VMX もしくは SVM 対応 CPU の場合は virt_type = kvm と設定することにより、仮想化部分のパフォーマンスが向上します。

```
compute# vi /etc/nova/nova-compute.conf

[libvirt]
...
virt_type = kvm
```

8.3　Nova-Compute サービスの再起動

　設定を反映させるため、Nova-Compute のサービスを再起動します。

```
compute# service nova-compute restart
```

8.4　controller ノードとの疎通確認

疎通確認は controller ノード上にて、admin 環境変数設定ファイルを読み込んで行います。

```
controller# source admin-openrc.sh
```

ホストリストの確認

controller ノードと compute ノードが相互に接続できているか確認します。もし、State が XXX なサービスがあった場合は、該当のサービスを service コマンドで起動してください。

```
controller# date -u
Wed Jul  1 08:59:20 UTC 2015          ← 現在時刻を確認

controller# nova-manage service list    ← Nova サービスステータスを確認
No handlers could be found for logger "oslo_config.cfg"
Binary           Host         Zone      Status    State Updated_At
nova-cert        controller   internal  enabled   :-)   2015-07-01 08:59:17
nova-consoleauth controller   internal  enabled   :-)   2015-07-01 08:59:17
nova-scheduler   controller   internal  enabled   :-)   2015-07-01 08:59:17
nova-conductor   controller   internal  enabled   :-)   2015-07-01 08:59:16
nova-compute     compute      nova      enabled   :-)   2015-07-01 08:59:18
```

※一覧に compute が表示されていれば問題ありません。

ハイパーバイザの確認

controller ノードより compute ノードのハイパーバイザが取得可能か確認します。

```
controller# nova hypervisor-list
+----+---------------------+-------+---------+
| ID | Hypervisor hostname | State | Status  |
+----+---------------------+-------+---------+
| 1  | compute             | up    | enabled |
+----+---------------------+-------+---------+
```

※ Hypervisor hostname 一覧に compute が表示されていれば問題ありません。

第9章 Neutronのインストール・設定（controllerノード）

9.1 データベース作成・確認

データベースの作成

MariaDBにデータベースneutronを作成します。

```
sql# mysql -u root -p << EOF
CREATE DATABASE neutron;
GRANT ALL PRIVILEGES ON neutron.* TO 'neutron'@'localhost' \
  IDENTIFIED BY 'password';
GRANT ALL PRIVILEGES ON neutron.* TO 'neutron'@'%' \
  IDENTIFIED BY 'password';
EOF
Enter password: ← MariaDBのrootパスワードpasswordを入力
```

データベースの確認

MariaDBにNeutronのデータベースが登録されたか確認します。

```
controller# mysql -h sql -u neutron -p
Enter password: ← MariaDBのneutronパスワードpasswordを入力
...

Type 'help;' or '\h' for help. Type '\c' to clear the current input statement.

MariaDB [(none)]> show databases;
+--------------------+
| Database           |
+--------------------+
| information_schema |
```

第 9 章　Neutron のインストール・設定（controller ノード）

```
| neutron            |
+--------------------+
2 rows in set (0.00 sec)
```

※ユーザー neutron でログイン可能でデータベースが閲覧可能なら問題ありません。

9.2　ユーザーとサービス、API エンドポイントの作成

以下コマンドで認証情報を読み込んだあと、サービスと API エンドポイントを設定します。

- 環境変数ファイルの読み込み

```
controller# source admin-openrc.sh
```

- neutron ユーザーの作成

```
controller# openstack user create --password-prompt neutron
User Password: password    #neutron ユーザーのパスワードを設定 (本例は password を設定)
Repeat User Password: password
+----------+----------------------------------+
| Field    | Value                            |
+----------+----------------------------------+
| email    | None                             |
| enabled  | True                             |
| id       | df3e33244d6548108edeaa7cc7b1789f |
| name     | neutron                          |
| username | neutron                          |
+----------+----------------------------------+
```

- neutron ユーザーを admin ロールに追加

```
controller# openstack role add --project service --user neutron admin
+-------+----------------------------------+
| Field | Value                            |
+-------+----------------------------------+
| id    | 9212e4ba1d07418a97fb4eaaaa275334 |
| name  | admin                            |
+-------+----------------------------------+
```

- neutron サービスの作成

```
controller# openstack service create --name neutron --description "OpenStack
Networking" network
+-------------+----------------------------------+
| Field       | Value                            |
+-------------+----------------------------------+
| description | OpenStack Networking             |
| enabled     | True                             |
| id          | efbab190a3b64d9db44725c3dc99b4c3 |
| name        | neutron                          |
| type        | network                          |
+-------------+----------------------------------+
```

- neutronサービスのAPIエンドポイントを作成

```
controller# openstack endpoint create \
--publicurl http://controller:9696 \
--adminurl http://controller:9696 \
--internalurl http://controller:9696 \
--region RegionOne network
+--------------+----------------------------------+
| Field        | Value                            |
+--------------+----------------------------------+
| adminurl     | http://controller:9696           |
| id           | fe4c5a9f82574cbeb2f17685bb0956c2 |
| internalurl  | http://controller:9696           |
| publicurl    | http://controller:9696           |
| region       | RegionOne                        |
| service_id   | efbab190a3b64d9db44725c3dc99b4c3 |
| service_name | neutron                          |
| service_type | network                          |
+--------------+----------------------------------+
```

9.3　パッケージのインストール

```
controller# apt-get update
controller# apt-get install -y neutron-server neutron-plugin-ml2
python-neutronclient
```

9.4　設定の変更

- Neutron Serverの設定

第 9 章　Neutron のインストール・設定（controller ノード）

```
controller# vi /etc/neutron/neutron.conf
[DEFAULT]
...
verbose = True
rpc_backend = rabbit              ←アンコメント
auth_strategy = keystone          ←アンコメント

core_plugin = ml2                 ←確認
service_plugins = router          ←追記
allow_overlapping_ips = True      ←変更

notify_nova_on_port_status_changes = True    ←アンコメント
notify_nova_on_port_data_changes = True      ←アンコメント
nova_url = http://controller:8774/v2         ←変更

[keystone_authtoken]（既存の設定はコメントアウトし、以下を追記）
...
auth_uri = http://controller:5000
auth_url = http://controller:35357
auth_plugin = password
project_domain_id = default
user_domain_id = default
project_name = service
username = neutron
password = password         ← neutron ユーザーのパスワード (9-2 で設定したもの)

[database]
#connection = sqlite:////var/lib/neutron/neutron.sqlite    ← 既存設定をコメントアウト
connection = mysql://neutron:password@sql/neutron          ← 追記

[nova]（以下末尾に追記）
...
auth_url = http://controller:35357
auth_plugin = password
project_domain_id = default
user_domain_id = default
region_name = RegionOne
project_name = service
username = nova
password = password         ← nova ユーザーのパスワード (6-2 で設定したもの)

（次ページに続きます...）
```

```
（前ページ/etc/neutron/neutron.conf の続き）

[oslo_messaging_rabbit]（以下追記）
...
# Deprecated group/name - [DEFAULT]/fake_rabbit
# fake_rabbit = false
rabbit_host = controller
rabbit_userid = openstack
```

```
rabbit_password = password
```

[keystone_authtoken] セクションは追記した設定以外は取り除くかコメントアウトしてください。

次のコマンドを実行して正しく設定を行ったか確認します。

```
controller# less /etc/neutron/neutron.conf | grep -v "^\s*$" | grep -v "^\s*#"
```

- ML2 プラグインの設定

```
controller# vi /etc/neutron/plugins/ml2/ml2_conf.ini

[ml2]
...
type_drivers = flat,vlan,gre,vxlan       ← 追記
tenant_network_types = gre               ← 追記
mechanism_drivers = openvswitch          ← 追記

[ml2_type_gre]
...
tunnel_id_ranges = 1:1000                ← 追記

[securitygroup]
...
enable_security_group = True             ← アンコメント
enable_ipset = True                      ← アンコメント
firewall_driver =
neutron.agent.linux.iptables_firewall.OVSHybridIptablesFirewallDriver   ← 追記
```

次のコマンドを実行して正しく設定を行ったか確認します。

```
controller# less /etc/neutron/plugins/ml2/ml2_conf.ini | grep -v "^\s*$" | grep -v "^\s*#"
```

9.5 設定の変更

Nova の設定ファイルに Neutron の設定を追記します。

```
controller# vi /etc/nova/nova.conf

[DEFAULT]
...
network_api_class = nova.network.neutronv2.api.API
security_group_api = neutron
linuxnet_interface_driver = nova.network.linux_net.LinuxOVSInterfaceDriver
```

```
firewall_driver = nova.virt.firewall.NoopFirewallDriver

[neutron]
url = http://controller:9696
auth_strategy = keystone
admin_auth_url = http://controller:35357/v2.0
admin_tenant_name = service
admin_username = neutron
admin_password = password         ← neutron ユーザーのパスワード（9-2 で設定したもの）
```

次のコマンドを実行して正しく設定を行ったか確認します。

```
controller# less /etc/nova/nova.conf | grep -v "^\s*$" | grep -v "^\s*#"
```

9.6　データベースの作成

コマンドを実行して、エラーが出ないで完了することを確認します。

```
controller# su -s /bin/sh -c "neutron-db-manage --config-file /etc/neutron/neutron.conf \
  --config-file /etc/neutron/plugins/ml2/ml2_conf.ini upgrade head" neutron
INFO  [alembic.migration] Context impl MySQLImpl.
INFO  [alembic.migration] Will assume non-transactional DDL.
...
INFO  [alembic.migration] Running upgrade 28a09af858a8 -> 20c469a5f920, add index for port
INFO  [alembic.migration] Running upgrade 20c469a5f920 -> kilo, kilo
```

9.7　ログの確認

インストールした Neutron Server のログを参照し、エラーが出ていないことを確認します。

```
controller# tailf /var/log/neutron/neutron-server.log
...
2015-07-03 11:32:50.570 8809 INFO neutron.service [-] Neutron service started, listening on 0.0.0.0:9696
2015-07-03 11:32:50.571 8809 INFO oslo_messaging._drivers.impl_rabbit [-] Connecting to AMQP server on controller:5672
2015-07-03 11:32:50.585 8809 INFO neutron.wsgi [-] (8809) wsgi starting up on http://0.0.0.0:9696/
2015-07-03 11:32:50.592 8809 INFO oslo_messaging._drivers.impl_rabbit [-] Connected to AMQP server on controller:5672
```

9.8 使用しないデータベースファイル削除

```
controller# rm /var/lib/neutron/neutron.sqlite
```

9.9 controllerノードのNeutronと関連サービスの再起動

設定を反映させるため、controllerノードの関連サービスを再起動します。

```
controller# service nova-api restart && service neutron-server restart
```

9.10 動作の確認

Neutron Serverの動作を確認するため、拡張機能一覧を表示するneutronコマンドを実行します。

```
controller:~# source admin-openrc.sh
controller:~# neutron ext-list
+-----------------------+---------------------------------------------+
| alias                 | name                                        |
+-----------------------+---------------------------------------------+
| security-group        | security-group                              |
| l3_agent_scheduler    | L3 Agent Scheduler                          |
| net-mtu               | Network MTU                                 |
| ext-gw-mode           | Neutron L3 Configurable external gateway mode |
| binding               | Port Binding                                |
| provider              | Provider Network                            |
| agent                 | agent                                       |
| quotas                | Quota management support                    |
| subnet_allocation     | Subnet Allocation                           |
| dhcp_agent_scheduler  | DHCP Agent Scheduler                        |
| l3-ha                 | HA Router extension                         |
| multi-provider        | Multi Provider Network                      |
| external-net          | Neutron external network                    |
| router                | Neutron L3 Router                           |
| allowed-address-pairs | Allowed Address Pairs                       |
| extraroute            | Neutron Extra Route                         |
| extra_dhcp_opt        | Neutron Extra DHCP opts                     |
| dvr                   | Distributed Virtual Router                  |
+-----------------------+---------------------------------------------+
```

第10章 Neutronのインストール・設定（networkノード）

10.1 パッケージのインストール

```
network# apt-get update
network# apt-get install -y neutron-plugin-ml2 neutron-plugin-openvswitch-agent \
neutron-l3-agent neutron-dhcp-agent neutron-metadata-agent
```

10.2 設定の変更

- Neutron の設定

```
network# vi /etc/neutron/neutron.conf

[DEFAULT]
...
verbose = True                      ← 変更
rpc_backend = rabbit                ← コメントアウトをはずす
auth_strategy = keystone            ← コメントアウトをはずす

core_plugin = ml2                   ← 確認
service_plugins = router            ← 追記
allow_overlapping_ips = True        ← 追記

[keystone_authtoken]（既存の設定はコメントアウトし、以下を追記）
...
auth_uri = http://controller:5000
auth_url = http://controller:35357
auth_plugin = password
project_domain_id = default
```

第 10 章　Neutron のインストール・設定（network ノード）

```
user_domain_id = default
project_name = service
username = neutron
password = password         ← neutron ユーザーのパスワード (9-2 で設定したもの)

[database]
# This line MUST be changed to actually run the plugin.
# Example:
#connection = sqlite:////var/lib/neutron/neutron.sqlite  ←コメントアウト

[oslo_messaging_rabbit]
...
# fake_rabbit = false
rabbit_host = controller
rabbit_userid = openstack
rabbit_password = password
```

本書の構成では、ネットワークノードの Neutron.conf にはデータベースの指定は不要です。
次のコマンドを実行して正しく設定を行ったか確認します。

```
network# less /etc/neutron/neutron.conf | grep -v "^\s*$" | grep -v "^\s*#"
```

- ML2 Plug-in の設定

```
network# vi /etc/neutron/plugins/ml2/ml2_conf.ini

[ml2]
...
type_drivers = flat,vlan,gre,vxlan       ← 追記
tenant_network_types = gre               ← 追記
mechanism_drivers = openvswitch          ← 追記

[ml2_type_flat]
...
flat_networks = external                 ← 追記

[ml2_type_gre]
...
tunnel_id_ranges = 1:1000                ← 追記

[securitygroup]
...
enable_security_group = True
enable_ipset = True
firewall_driver =
neutron.agent.linux.iptables_firewall.OVSHybridIptablesFirewallDriver

[agent]
tunnel_types = gre                       ← 追記
```

```
[ovs]
local_ip = 192.168.0.102              ← 追記 (network ノードの Internal 側)
enable_tunneling = True               ← 追記
bridge_mappings = external:br-ex      ← 追記
```

次のコマンドを実行して正しく設定を行ったか確認します。

```
network# less /etc/neutron/plugins/ml2/ml2_conf.ini | grep -v "^\s*$" | grep -v "^\s*#"
```

- Layer-3 (L3) agent の設定

```
network# vi /etc/neutron/l3_agent.ini

[DEFAULT]
interface_driver = neutron.agent.linux.interface.OVSInterfaceDriver   ← アンコメント
router_delete_namespaces = True        ← 変更
external_network_bridge = br-ex        ← アンコメント
verbose = True                         ← 追記
```

次のコマンドを実行して正しく設定を行ったか確認します。

```
network# less /etc/neutron/l3_agent.ini | grep -v "^\s*$" | grep -v "^\s*#"
```

- DHCP agent の設定

```
network# vi /etc/neutron/dhcp_agent.ini

[DEFAULT]
...
interface_driver = neutron.agent.linux.interface.OVSInterfaceDriver   ← アンコメント
dhcp_driver = neutron.agent.linux.dhcp.Dnsmasq                        ← アンコメント
dhcp_delete_namespaces = True                                         ← 変更
dnsmasq_config_file = /etc/neutron/dnsmasq-neutron.conf    ← 追記
verbose = True        ← 追記
```

次のコマンドを実行して正しく設定を行ったか確認します。

```
network# less /etc/neutron/dhcp_agent.ini | grep -v "^\s*$" | grep -v "^\s*#"
```

- DHCP オプションで MTU の設定

dnsmasq-neutron.conf ファイルを新規作成して、DHCP オプションを設定します。

第 10 章　Neutron のインストール・設定（network ノード）

```
network# vi /etc/neutron/dnsmasq-neutron.conf
dhcp-option-force=26,1454
```

- Metadata agent の設定

```
network# vi /etc/neutron/metadata_agent.ini

[DEFAULT]
...
auth_url = http://localhost:5000/v2.0         ← コメントアウト
admin_tenant_name = %SERVICE_TENANT_NAME%     ← コメントアウト
admin_user = %SERVICE_USER%                   ← コメントアウト
admin_password = %SERVICE_PASSWORD%           ← コメントアウト
auth_region = RegionOne                       ← 確認
（以下追記）
verbose = True
auth_uri = http://controller:5000
auth_url = http://controller:35357
auth_plugin = password
project_domain_id = default
user_domain_id = default
project_name = service
username = neutron
password = password          ← neutron ユーザーのパスワード (9-2 で設定したもの)
nova_metadata_ip = controller
metadata_proxy_shared_secret = password
```

metadata_proxy_shared_secret はコマンドを実行して生成したハッシュ値を設定することを推奨します。

[実行例]

```
# openssl rand -hex 10
```

次のコマンドを実行して正しく設定を行ったか確認します。

```
network# less /etc/neutron/metadata_agent.ini | grep -v "^\s*$" | grep -v "^\s*#"
```

10.3　設定の変更

controller ノードの Nova の設定ファイルに追記します。

```
controller# vi /etc/nova/nova.conf

[neutron]
...
```

```
service_metadata_proxy = True
metadata_proxy_shared_secret = password   ← ハッシュ値 (9-2「Metadata agent」に設定したも
のと同じもの)
```

次のコマンドを実行して正しく設定を行ったか確認します。

```
controller# less /etc/nova/nova.conf | grep -v "^\s*$" | grep -v "^\s*#"
```

controller ノードの nova-api サービスを再起動します。

```
controller# service nova-api restart
```

10.4　network ノードの Open vSwitch サービスの再起動

OpenStack のネットワークサービス設定を反映させるため、network ノードで Open vSwitch のサービスを再起動します。

```
network# service openvswitch-switch restart
```

10.5　ブリッジデバイス設定

内部通信用と外部通信用のブリッジを作成して外部通信用ブリッジに共有ネットワークデバイスを接続します。

```
network# ovs-vsctl add-br br-ex ; ovs-vsctl add-port br-ex eth0
```

注意: このコマンドを実行すると network ノードへの SSH 接続が切断されます。SSH 接続ではなく、HP iLo、Dell iDRAC などのリモートコンソールやサーバーコンソール上でコマンドを実行することを推奨します。

10.6　サービスの再起動

設定を反映するために、関連サービスを再起動します。

```
network# service neutron-plugin-openvswitch-agent restart
network# service neutron-l3-agent restart
network# service neutron-dhcp-agent restart
```

第 10 章 Neutron のインストール・設定（network ノード）

```
network# service neutron-metadata-agent restart
```

10.7　ブリッジデバイス設定確認

ブリッジの作成・設定を確認します。

ブリッジの確認

```
network# ovs-vsctl list-br
br-ex
br-int
br-tun
```

※ add-br したブリッジが表示されていれば問題ありません。

外部接続用ブリッジと共有ネットワークデバイスの接続確認

```
network# ovs-vsctl list-ports br-ex
eth0
phy-br-ex
```

※ add-port で設定したネットワークデバイスが表示されていれば問題ありません。

10.8　ネットワークインターフェースの設定変更

Management 側に接続された NIC を使って、仮想 NIC(br-ex) を作成します。

```
network# vi /etc/network/interfaces

# This file describes the network interfaces available on your system
# and how to activate them. For more information, see interfaces(5).

# The loopback network interface
auto lo
iface lo inet loopback

auto eth0
iface eth0 inet manual                          ← 既存設定を変更
        up ip link set dev $IFACE up            ← 既存設定を変更
        down ip link set dev $IFACE down        ← 既存設定を変更

auto br-ex                                      ← 追記
iface br-ex inet static                         ← 追記
```

```
        address 10.0.0.102                    ← 追記
        netmask 255.255.255.0                 ← 追記
        gateway 10.0.0.1                      ← 追記
        dns-nameservers 10.0.0.1              ← 追記
auto eth1
iface eth1 inet static
        address 192.168.0.102
        netmask 255.255.255.0
```

10.9　networkノードの再起動

インターフェース設定を適用するために、システムを再起動します。

```
network# reboot
```

10.10　ブリッジの設定を確認

各種ブリッジが正常に設定されていることを確認します。

```
network# ip a |grep 'LOOPBACK\|BROADCAST\|inet'
1: lo: <LOOPBACK,UP,LOWER_UP> mtu 65536 qdisc noqueue state UNKNOWN group default
    inet 127.0.0.1/8 scope host lo
    inet6 ::1/128 scope host
2: eth0: <BROADCAST,MULTICAST,UP,LOWER_UP> mtu 1500 qdisc mq master ovs-system state UP group default qlen 1000
    inet6 fe80::20c:29ff:fe61:9417/64 scope link
3: eth1: <BROADCAST,MULTICAST,UP,LOWER_UP> mtu 1500 qdisc mq state UP group default qlen 1000
    inet 192.168.14.102/24 brd 192.168.14.255 scope global eth1
    inet6 fe80::20c:29ff:fe61:9421/64 scope link
4: ovs-system: <BROADCAST,MULTICAST> mtu 1500 qdisc noop state DOWN group default
5: br-ex: <BROADCAST,MULTICAST,UP,LOWER_UP> mtu 1500 qdisc noqueue state UNKNOWN group default
    inet 172.17.14.102/24 brd 172.17.14.255 scope global br-ex
    inet6 fe80::20c:29ff:fe61:9417/64 scope link
6: br-int: <BROADCAST,MULTICAST> mtu 1500 qdisc noop state DOWN group default
7: br-tun: <BROADCAST,MULTICAST> mtu 1500 qdisc noop state DOWN group default
```

10.11　Neutronサービスの動作確認

構築したNeutronのエージェントが正しく認識され、稼働していることを確認します。

第 10 章　Neutron のインストール・設定（network ノード）

```
controller# source admin-openrc.sh
controller# neutron agent-list -c host -c alive -c binary
+---------+-------+-------------------------+
| host    | alive | binary                  |
+---------+-------+-------------------------+
| network | :-)   | neutron-dhcp-agent      |
| network | :-)   | neutron-l3-agent        |
| network | :-)   | neutron-metadata-agent  |
| network | :-)   | neutron-openvswitch-agent |
+---------+-------+-------------------------+
```

第11章 Neutronのインストール・設定（computeノード）

11.1 パッケージのインストール

```
compute# apt-get update
compute# apt-get install -y neutron-plugin-ml2 neutron-plugin-openvswitch-agent
```

11.2 設定の変更

- Neutronの設定

```
compute# vi /etc/neutron/neutron.conf

[DEFAULT]
...
verbose = True
rpc_backend = rabbit                    ← アンコメント
auth_strategy = keystone                ← アンコメント

core_plugin = ml2                       ← 確認
service_plugins = router                ← 追記
allow_overlapping_ips = True            ← 追記

[keystone_authtoken]（既存の設定はコメントアウトし、以下を追記）
...
auth_uri = http://controller:5000
auth_url = http://controller:35357
auth_plugin = password
project_domain_id = default
user_domain_id = default
```

第 11 章 Neutron のインストール・設定（compute ノード）

```
project_name = service
username = neutron
password = password          ← neutron ユーザーのパスワード (9-2 で設定したもの)

[database]
# This line MUST be changed to actually run the plugin.
# Example:
# connection = sqlite:////var/lib/neutron/neutron.sqlite   ← コメントアウト

[oslo_messaging_rabbit]
...
# fake_rabbit = false
rabbit_host = controller           ← 追記
rabbit_userid = openstack          ← 追記
rabbit_password = password         ← 追記
```

本書の構成では、compute ノードの Neutron.conf にはデータベースの指定は不要です。

次のコマンドを実行して正しく設定を行ったか確認します。

```
compute# less /etc/neutron/neutron.conf | grep -v "^\s*$" | grep -v "^\s*#"
```

- ML2 Plug-in の設定

```
compute# vi /etc/neutron/plugins/ml2/ml2_conf.ini

[ml2]
type_drivers = flat,vlan,gre,vxlan    ← 追記
tenant_network_types = gre            ← 追記
mechanism_drivers = openvswitch       ← 追記

[ml2_type_gre]
tunnel_id_ranges = 1:1000             ← 追記

[securitygroup]
enable_security_group = True          ← 追記
enable_ipset = True                   ← 追記
firewall_driver =
neutron.agent.linux.iptables_firewall.OVSHybridIptablesFirewallDriver    ← 追記

[ovs]                                 ← 追記
local_ip = 192.168.0.103              ← 追記 (compute ノードの Internal 側)
enable_tunneling = True               ← 追記

[agent]                               ← 追記
tunnel_types = gre                    ← 追記
```

次のコマンドを実行して正しく設定を行ったか確認します。

```
compute# less /etc/neutron/plugins/ml2/ml2_conf.ini | grep -v "^\s*$" | grep -v
"^\s*#"
```

11.3　compute ノードの Open vSwitch サービスの再起動

設定を反映させるため、compute ノードの Open vSwitch のサービスを再起動します。

```
compute# service openvswitch-switch restart
```

11.4　compute ノードのネットワーク設定

デフォルトでは Compute はレガシーなネットワークを利用します。Neutron を利用するように設定を変更します。

```
compute# vi /etc/nova/nova.conf

[DEFAULT]
...
network_api_class = nova.network.neutronv2.api.API
security_group_api = neutron
linuxnet_interface_driver = nova.network.linux_net.LinuxOVSInterfaceDriver
firewall_driver = nova.virt.firewall.NoopFirewallDriver

[neutron]
url = http://controller:9696
auth_strategy = keystone
admin_auth_url = http://controller:35357/v2.0
admin_tenant_name = service
admin_username = neutron
admin_password = password         ← neutron ユーザーのパスワード (9-2 で設定したもの)
```

次のコマンドを実行して正しく設定を行ったか確認します。

```
compute# less /etc/nova/nova.conf | grep -v "^\s*$" | grep -v "^\s*#"
```

11.5　compute ノードの Neutron と関連サービスを再起動

ネットワーク設定を反映させるため、compute1 ノードの Neutron と関連のサービスを再起動

します。

```
compute# service nova-compute restart && service neutron-plugin-openvswitch-agent restart
```

11.6　ログの確認

```
compute# grep "ERROR\|WARNING" /var/log/neutron/*
```

※何も表示されなければ問題ありません。RabbitMQ の接続確立に時間がかかり、その間「AMQP server on 127.0.0.1:5672 is unreachable」というエラーが出力される場合があります。

11.7　Neutron サービスの動作の確認

構築した Neutron のエージェントが正しく認識され、稼働していることを確認します。

※コンピュートが追加され、正常に稼働していることが確認できれば問題ありません。

第12章 仮想ネットワーク設定 (controllerノード)

12.1 外部接続ネットワークの設定

admin環境変数ファイルの読み込み

外部接続用ネットワーク作成するためにadmin環境変数を読み込みます。

```
controller# source admin-openrc.sh
```

外部ネットワーク作成

ext-netという名前で外部用ネットワークを作成します。

```
controller# neutron net-create ext-net --router:external \
--provider:physical_network external --provider:network_type flat
Created a new network:
+---------------------------+--------------------------------------+
| Field                     | Value                                |
+---------------------------+--------------------------------------+
| admin_state_up            | True                                 |
| id                        | 37bc47f1-e8b9-4d43-ab3b-c1926b2b42b3 |
| mtu                       | 0                                    |
| name                      | ext-net                              |
| provider:network_type     | flat                                 |
| provider:physical_network | external                             |
| provider:segmentation_id  |                                      |
| router:external           | True                                 |
| shared                    | False                                |
| status                    | ACTIVE                               |
| subnets                   |                                      |
| tenant_id                 | 218010a87fe5477bba7f5e25c8211614     |
```

```
+--------------------------------+------------------------------------+
```

外部ネットワーク用サブネットを作成

ext-subnet という名前で外部ネットワーク用サブネットを作成します。

```
controller# neutron subnet-create ext-net --name ext-subnet \
  --allocation-pool start=10.0.0.200,end=10.0.0.250 \
  --disable-dhcp --gateway 10.0.0.1 10.0.0.0/24
Created a new subnet:
+-------------------+------------------------------------------------+
| Field             | Value                                          |
+-------------------+------------------------------------------------+
| allocation_pools  | {"start": "10.0.0.200", "end": "10.0.0.250"}   |
| cidr              | 10.0.0.0/24                                    |
| dns_nameservers   |                                                |
| enable_dhcp       | False                                          |
| gateway_ip        | 10.0.0.1                                       |
| host_routes       |                                                |
| id                | 406b93cb-3f55-49b3-8ce4-74259ac00526           |
| ip_version        | 4                                              |
| ipv6_address_mode |                                                |
| ipv6_ra_mode      |                                                |
| name              | ext-subnet                                     |
| network_id        | daf2a1a8-615d-4105-bfb4-60a8380350ef           |
| subnetpool_id     |                                                |
| tenant_id         | 218010a87fe5477bba7f5e25c8211614               |
+-------------------+------------------------------------------------+
```

12.2 インスタンス用ネットワークの設定

demo 環境変数ファイルの読み込み

インスタンス用ネットワーク作成するために demo 環境変数を読み込みます。

```
controller# source demo-openrc.sh
```

インスタンス用ネットワークの作成

demo-net という名前でインスタンス用ネットワークを作成します。

```
controller# neutron net-create demo-net
Created a new network:
+----------------+------------------------------------+
| Field          | Value                              |
+----------------+------------------------------------+
```

```
| admin_state_up  | True                                 |
| id              | de074cba-bcad-47ca-ab6d-058a974000b5 |
| mtu             | 0                                    |
| name            | demo-net                             |
| router:external | False                                |
| shared          | False                                |
| status          | ACTIVE                               |
| subnets         |                                      |
| tenant_id       | 3ed2437abf474a37b305338666d9fafa     |
+-----------------+--------------------------------------+
```

11-2-3 インスタンス用ネットワークサブネットを作成

demo-subnet という名前でインスタンス用ネットワークサブネットを作成します。

```
controller# neutron subnet-create demo-net 192.168.0.0/24 \
--name demo-subnet --gateway 192.168.0.1 --dns-nameserver 8.8.8.8
Created a new subnet:
+-------------------+----------------------------------------------------+
| Field             | Value                                              |
+-------------------+----------------------------------------------------+
| allocation_pools  | {"start": "192.168.0.2", "end": "192.168.0.254"}   |
| cidr              | 192.168.0.0/24                                     |
| dns_nameservers   |                                                    |
| enable_dhcp       | True                                               |
| gateway_ip        | 192.168.0.1                                        |
| host_routes       |                                                    |
| id                | 92b87319-b287-4bcc-988c-003215c1580a               |
| ip_version        | 4                                                  |
| ipv6_address_mode |                                                    |
| ipv6_ra_mode      |                                                    |
| name              | demo-subnet                                        |
| network_id        | 6be6a7ef-ef68-4c84-9b3f-0a6e41aee52b               |
| subnetpool_id     |                                                    |
| tenant_id         | 3ed2437abf474a37b305338666d9fafa                   |
+-------------------+----------------------------------------------------+
```

12.3　仮想ネットワークルーターの設定

仮想ネットワークルーターを作成して外部接続用ネットワークとインスタンス用ネットワークをルーターに接続し、双方でデータのやり取りを行えるようにします。

demo-router を作成

仮想ネットワークルーターを作成します。

第 12 章　仮想ネットワーク設定（controller ノード）

```
controller# neutron router-create demo-router
Created a new router:
+-----------------------+--------------------------------------+
| Field                 | Value                                |
+-----------------------+--------------------------------------+
| admin_state_up        | True                                 |
| external_gateway_info |                                      |
| id                    | 8dea222a-cf31-4de0-a946-891026c21364 |
| name                  | demo-router                          |
| routes                |                                      |
| status                | ACTIVE                               |
| tenant_id             | 3ed2437abf474a37b305338666d9fafa     |
+-----------------------+--------------------------------------+
```

demo-router にサブネットを追加

仮想ネットワークルーターにインスタンス用ネットワークを接続します。

```
controller# neutron router-interface-add demo-router demo-subnet
Added interface a66a184a-55b3-49d8-bbbf-3bbf2fe32de2 to router demo-router.
```

demo-router にゲートウェイを追加

仮想ネットワークルーターに外部ネットワークを接続します。

```
controller# neutron router-gateway-set demo-router ext-net
Set gateway for router demo-router
```

第13章 仮想ネットワーク設定確認（networkノード）

13.1 仮想ネットワークルーターの確認

以下コマンドで仮想ネットワークルーターが作成されているか確認します。

```
network# ip netns
qdhcp-ed07c38c-8609-43d8-ae02-582f9f202a3e
qrouter-7c1ca8eb-eaa0-4a68-843d-daca30824693
```

※ qrouter~~ という名前の行が表示されていれば問題ありません。

13.2 仮想ルーターのネームスペースのIPアドレスを確認

仮想ルーターと外部用ネットワークの接続を確認します。

```
network# ip netns exec `ip netns | grep qrouter` ip addr
1: lo: <LOOPBACK,UP,LOWER_UP> mtu 65536 qdisc noqueue state UNKNOWN group default
    link/loopback 00:00:00:00:00:00 brd 00:00:00:00:00:00
    inet 127.0.0.1/8 scope host lo
       valid_lft forever preferred_lft forever
    inet6 ::1/128 scope host
       valid_lft forever preferred_lft forever
13: qr-65249869-77: <BROADCAST,UP,LOWER_UP> mtu 1500 qdisc noqueue state UNKNOWN group default
    link/ether fa:16:3e:d0:df:2c brd ff:ff:ff:ff:ff:ff
    inet 192.168.0.1/24 brd 192.168.0.255 scope global qr-65249869-77
       valid_lft forever preferred_lft forever
    inet6 fe80::f816:3eff:fed0:df2c/64 scope link
```

```
        valid_lft forever preferred_lft forever
14: qg-bd7c5797-3f: <BROADCAST,UP,LOWER_UP> mtu 1500 qdisc noqueue state UNKNOWN
group default
    link/ether fa:16:3e:35:80:8f brd ff:ff:ff:ff:ff:ff
    inet 10.0.0.200/24 brd 10.0.0.255 scope global qg-bd7c5797-3f
        valid_lft forever preferred_lft forever
    inet6 fe80::f816:3eff:fe35:808f/64 scope link
        valid_lft forever preferred_lft forever
```

※外部アドレス（この環境では 10.0.0.0/24）のアドレスを確認します。

13.3　仮想ゲートウェイの疎通を確認

仮想ルーターの外から仮想ルーターに対して疎通可能かを確認します。

```
network# ping 10.0.0.200
PING 10.0.0.200 (10.0.0.200) 56(84) bytes of data.
64 bytes from 10.0.0.200: icmp_seq=1 ttl=64 time=1.17 ms
64 bytes from 10.0.0.200: icmp_seq=2 ttl=64 time=0.074 ms
64 bytes from 10.0.0.200: icmp_seq=3 ttl=64 time=0.061 ms
64 bytes from 10.0.0.200: icmp_seq=4 ttl=64 time=0.076 ms
```

仮想ルーターからゲートウェイ、外部ネットワークにアクセス可能か確認します。

```
network# ip netns exec `ip netns | grep qrouter` ping 10.0.0.1
network# ip netns exec `ip netns | grep qrouter` ping virtualtech.jp
```

※応答が返ってくれば問題ありません。各ノードから Ping コマンドによる疎通確認を実行しましょう。

13.4　インスタンスの起動確認

　controller、network、compute ノードの最低限の構成ができあがったので、ここで OpenStack 環境がうまく動作しているか確認しましょう。まずはコマンドを使ってインスタンスを起動するために必要な情報を集める所から始めます。環境設定ファイルを読み込んで、各コマンドを実行し、情報を集めてください。

```
controller# source demo-openrc.sh

controller# glance image-list
+--------------------------------------+----------------------+
| ID                                   | Name                 |
+--------------------------------------+----------------------+
```

13.4 インスタンスの起動確認

```
| 572104d5-a901-4432-be41-17e37901a5f7 | cirros-0.3.4-x86_64 |
+--------------------------------------+---------------------+

controller# neutron net-list -c name -c id
+----------+--------------------------------------+
| name     | id                                   |
+----------+--------------------------------------+
| demo-net | 858c0fe5-ea00-4426-aa4e-f2a2484b2471 |
| ext-net  | e0cbae79-2973-4870-93de-09dfbd3e76e4 |
+----------+--------------------------------------+

controller# nova secgroup-list
+--------------------------------------+---------+------------------------+
| Id                                   | Name    | Description            |
+--------------------------------------+---------+------------------------+
| a66e2962-312f-45a4-bfd3-f86ec69c5582 | default | Default security group |
+--------------------------------------+---------+------------------------+

controller# nova flavor-list
+----+-----------+-----------+------+-----------+------+-------+-------------+-----------+
| ID | Name      | Memory_MB | Disk | Ephemeral | Swap | VCPUs | RXTX_Factor | Is_Public |
+----+-----------+-----------+------+-----------+------+-------+-------------+-----------+
| 1  | m1.tiny   | 512       | 1    | 0         |      | 1     | 1.0         | True      |
| 2  | m1.small  | 2048      | 20   | 0         |      | 1     | 1.0         | True      |
| 3  | m1.medium | 4096      | 40   | 0         |      | 2     | 1.0         | True      |
| 4  | m1.large  | 8192      | 80   | 0         |      | 4     | 1.0         | True      |
| 5  | m1.xlarge | 16384     | 160  | 0         |      | 8     | 1.0         | True      |
+----+-----------+-----------+------+-----------+------+-------+-------------+-----------+
```

nova boot コマンドを使って、インスタンスを起動します。正常に起動したら nova delete コマンドでインスタンスを削除してください。

```
controller# nova boot --flavor m1.tiny --image "cirros-0.3.4-x86_64" --nic
net-id=858c0fe5-ea00-4426-aa4e-f2a2484b2471 --security-group
a66e2962-312f-45a4-bfd3-f86ec69c5582 vm1
(インスタンスを起動)

controller# watch nova list
(インスタンス一覧を表示)
+--------------------------------------+------+--------+------------+-------------+----------------------+
| ID                                   | Name | Status | Task State | Power State | Networks             |
+--------------------------------------+------+--------+------------+-------------+----------------------+
| 5eddf2a7-0287-46ef-b656-18ff51f1c605 | vm1  | ACTIVE | -          | Running     | demo-net=192.168.0.6 |
+--------------------------------------+------+--------+------------+-------------+----------------------+
```

第 13 章　仮想ネットワーク設定確認（network ノード）

```
# grep "ERROR\|WARNING" /var/log/rabbitmq/*.log
# grep "ERROR\|WARNING" /var/log/openvswitch/*
# grep "ERROR\|WARNING" /var/log/neutron/*
# grep "ERROR\|WARNING" /var/log/nova/*
(各ノードの関連サービスでエラーが出ていないことを確認)

controller# nova delete vm1
Request to delete server vm1 has been accepted.
(起動したインスタンスを削除)
```

第14章 Cinderのインストール（controllerノード）

14.1 データベース作成・確認

データベースの作成

MariaDBのデータベースにCinderのデータベースを作成します。

```
sql# mysql -u root -p << EOF
CREATE DATABASE cinder;
GRANT ALL PRIVILEGES ON cinder.* TO 'cinder'@'localhost' \
  IDENTIFIED BY 'password';
GRANT ALL PRIVILEGES ON cinder.* TO 'cinder'@'%' \
  IDENTIFIED BY 'password';
EOF
Enter password: ← MariaDBのrootパスワードpasswordを入力
```

データベースの確認

MariaDBにCinderのデータベースが登録されたか確認します。

```
controller# mysql -h sql -u cinder -p
Enter password: ← MariaDBのcinderパスワードpasswordを入力
...
Type 'help;' or '\h' for help. Type '\c' to clear the current input statement.

MariaDB [(none)]> show databases;
+--------------------+
| Database           |
+--------------------+
| information_schema |
| cinder             |
```

第 14 章　Cinder のインストール（controller ノード）

```
+-------------------+
2 rows in set (0.00 sec)
```

※ユーザー cinder でログイン可能でデータベースの閲覧が可能なら問題ありません。

14.2　ユーザーとサービス、API エンドポイントの作成

以下コマンドで認証情報を読み込んだあと、サービスと API エンドポイントを設定します。

- 環境変数ファイルの読み込み

```
controller# source admin-openrc.sh
```

- cinder ユーザーの作成

```
controller# openstack user create --password-prompt cinder
User Password: password   #cinder ユーザーのパスワードを設定 (本例は password を設定)
Repeat User Password: password
+----------+----------------------------------+
| Field    | Value                            |
+----------+----------------------------------+
| email    | None                             |
| enabled  | True                             |
| id       | f63d06a517eb484f919276bdba5b9567 |
| name     | cinder                           |
| username | cinder                           |
+----------+----------------------------------+
```

- cinder ユーザーを admin ロールに追加

```
controller# openstack role add --project service --user cinder admin
+-------+----------------------------------+
| Field | Value                            |
+-------+----------------------------------+
| id    | 9212e4ba1d07418a97fb4eaaaa275334 |
| name  | admin                            |
+-------+----------------------------------+
```

- cinder サービスの作成

14.2 ユーザーとサービス、API エンドポイントの作成

```
controller# openstack service create --name cinder \
--description "OpenStack Block Storage" volume
+-------------+----------------------------------+
| Field       | Value                            |
+-------------+----------------------------------+
| description | OpenStack Block Storage          |
| enabled     | True                             |
| id          | 546be85aaa664c71bc35add65c98a224 |
| name        | cinder                           |
| type        | volume                           |
+-------------+----------------------------------+

controller# openstack service create --name cinderv2 \
--description "OpenStack Block Storage" volumev2
+-------------+----------------------------------+
| Field       | Value                            |
+-------------+----------------------------------+
| description | OpenStack Block Storage          |
| enabled     | True                             |
| id          | 77ac9832fdcd485b928c1183270277757|
| name        | cinderv2                         |
| type        | volumev2                         |
+-------------+----------------------------------+
```

- Block Storage サービスの API エンドポイントを作成

```
controller# openstack endpoint create \
--publicurl http://controller:8776/v2/%\(tenant_id\)s \
--internalurl http://controller:8776/v2/%\(tenant_id\)s \
--adminurl http://controller:8776/v2/%\(tenant_id\)s \
--region RegionOne volume
+--------------+-----------------------------------------+
| Field        | Value                                   |
+--------------+-----------------------------------------+
| adminurl     | http://controller:8776/v2/%(tenant_id)s |
| id           | 29b1496fed5145bfb6fafc90c265010f        |
| internalurl  | http://controller:8776/v2/%(tenant_id)s |
| publicurl    | http://controller:8776/v2/%(tenant_id)s |
| region       | RegionOne                               |
| service_id   | 546be85aaa664c71bc35add65c98a224        |
| service_name | cinder                                  |
| service_type | volume                                  |
+--------------+-----------------------------------------+

controller# openstack endpoint create \
--publicurl http://controller:8776/v2/%\(tenant_id\)s \
--internalurl http://controller:8776/v2/%\(tenant_id\)s \
--adminurl http://controller:8776/v2/%\(tenant_id\)s \
--region RegionOne volumev2
+--------------+-----------------------------------------+
| Field        | Value                                   |
```

第 14 章　Cinder のインストール（controller ノード）

```
+--------------+------------------------------------------+
| adminurl     | http://controller:8776/v2/%(tenant_id)s  |
| id           | 18777ee389dd4bfcbf5cc78920ce9f44         |
| internalurl  | http://controller:8776/v2/%(tenant_id)s  |
| publicurl    | http://controller:8776/v2/%(tenant_id)s  |
| region       | RegionOne                                |
| service_id   | 77ac9832fdcd485b928c118327027757         |
| service_name | cinderv2                                 |
| service_type | volumev2                                 |
+--------------+------------------------------------------+
```

14.3　パッケージのインストール

　本書では Block Storage コントローラーと Block Storage ボリュームコンポーネントを 1 台のマシンで構築するため、両方の役割をインストールします。

```
controller# apt-get update
controller# apt-get install -y lvm2 cinder-api cinder-scheduler cinder-volume python-mysqldb python-cinderclient
```

14.4　設定の変更

```
controller# vi /etc/cinder/cinder.conf

[DEFAULT]
...
verbose = True              ← 確認
auth_strategy = keystone    ← 確認
rpc_backend = rabbit        ← 追記

my_ip = 10.0.0.101     #controller ノード
enabled_backends = lvm
glance_host = controller

[oslo_messaging_rabbit]
rabbit_host = controller
rabbit_userid = openstack
rabbit_password = password

[oslo_concurrency]
lock_path = /var/lock/cinder

[database]
connection = mysql://cinder:password@sql/cinder
```

14.8 イメージ格納用ボリュームの作成

```
[keystone_authtoken]
auth_uri = http://controller:5000
auth_url = http://controller:35357
auth_plugin = password
project_domain_id = default
user_domain_id = default
project_name = service
username = cinder
password = password         ← cinder ユーザーのパスワード (13-2 で設定したもの)

[lvm]
volume_driver = cinder.volume.drivers.lvm.LVMVolumeDriver
volume_group = cinder-volumes
iscsi_protocol = iscsi
iscsi_helper = tgtadm
```

次のコマンドを実行して正しく設定を行ったか確認します。

```
controller# less /etc/cinder/cinder.conf | grep -v "^\s*$" | grep -v "^\s*#"
```

14.5　データベースに表を作成

```
controller# su -s /bin/sh -c "cinder-manage db sync" cinder
```

14.6　Cinderサービスの再起動

設定を反映させるために、Cinder のサービスを再起動します。

```
controller# service cinder-scheduler restart && service cinder-api restart
```

14.7　使用しないデータベースファイルの削除

```
controller# rm /var/lib/cinder/cinder.sqlite
```

14.8　イメージ格納用ボリュームの作成

イメージ格納用ボリュームを設定するために物理ボリュームの設定、ボリューム作成を行います。

物理ボリュームを追加

本例では controller ノードにハードディスクを追加して、そのボリュームを Cinder 用ボリュームとして使います。controller ノードをいったんシャットダウンしてからハードディスクを増設し、再起動してください。新しく増設したディスクは dmesg コマンドや fdisk -l コマンドなどを使って確認できます。

仮想マシンにハードディスクを増設した場合は /dev/vdb などのようにデバイス名が異なる場合があります。

物理ボリュームを設定

以下のコマンドで物理ボリュームを作成します。

- LVM 物理ボリュームの作成

```
controller# pvcreate /dev/sdb
  Physical volume "/dev/sdb" successfully created
```

- LVM ボリュームグループの作成

```
controller# vgcreate cinder-volumes /dev/sdb
  Volume group "cinder-volumes" successfully created
```

Cinder-Volume サービスの再起動

Cinder ストレージの設定を反映させるために、Cinder-Volume のサービスを再起動します。

```
controller# service cinder-volume restart && service tgt restart
```

admin 環境変数設定ファイルの読み込み

Block Storage クライアントで API 2.0 でアクセスするように環境変数設定ファイルを書き換えます。

```
controller# echo "export OS_VOLUME_API_VERSION=2" | tee -a admin-openrc.sh demo-openrc.sh
```

インスタンス格納用ボリュームを作成するために、admin 環境変数を読み込みます。

14.8 イメージ格納用ボリュームの作成

```
controller# source admin-openrc.sh
```

ボリュームの作成

以下のコマンドでインスタンス格納用ボリュームを作成します。

```
controller# cinder create --display-name testvolume01 1
+---------------------------------------+--------------------------------------+
|               Property                |                Value                 |
+---------------------------------------+--------------------------------------+
|              attachments              |                  []                  |
|           availability_zone           |                 nova                 |
|                bootable               |                false                 |
|           consistencygroup_id         |                 None                 |
|              created_at               |      2015-06-25T08:41:05.000000      |
|              description              |                 None                 |
|               encrypted               |                False                 |
|                  id                   | 776ed580-780e-431b-8d5a-f4d883b98884 |
|                metadata               |                  {}                  |
|              multiattach              |                False                 |
|                 name                  |             testvolume01             |
|         os-vol-host-attr:host         |                 None                 |
|     os-vol-mig-status-attr:migstat    |                 None                 |
|     os-vol-mig-status-attr:name_id    |                 None                 |
|      os-vol-tenant-attr:tenant_id     |   218010a87fe5477bba7f5e25c8211614   |
|    os-volume-replication:driver_data  |                 None                 |
|  os-volume-replication:extended_status|                 None                 |
|          replication_status           |               disabled               |
|                 size                  |                  1                   |
|             snapshot_id               |                 None                 |
|             source_volid              |                 None                 |
|                status                 |               creating               |
|                user_id                |   9caffb5dc1d749c5b3e9493139fe8598   |
|              volume_type              |                 None                 |
+---------------------------------------+--------------------------------------+
```

作成ボリュームの確認

以下のコマンドで作成したボリュームを確認します。

```
controller# cinder list
+--------------------------------------+-----------+--------------+------+-------------
|                  ID                  |  Status   |     Name     | Size | Volume
 Type | Bootable | Attached to |
+--------------------------------------+-----------+--------------+------+-------------
| 776ed580-780e-431b-8d5a-f4d883b98884 | available | testvolume01 |  1   |     None
| false    |          |
+--------------------------------------+-----------+--------------+------+-------------

controller# cinder delete testvolume01
```

第14章 Cinder のインストール (controller ノード)

(作成したテストボリュームの削除)

※一覧にコマンドを実行して登録したボリュームが表示されて、ステータスが available となっていれば問題ありません。

第15章 Dashboardインストール・確認(controllerノード)

クライアントマシンからブラウザーで OpenStack 環境を操作可能な Web インターフェースをインストールします。

15.1 パッケージのインストール

controller ノードに Dashboard をインストールします。

```
controller# apt-get update
controller# apt-get install -y openstack-dashboard
```

15.2 Dashboardの設定の変更

インストールした Dashboard の設定を変更します。

```
controller# vi /etc/openstack-dashboard/local_settings.py

...
OPENSTACK_HOST = "controller"        ← 変更
ALLOWED_HOSTS = '*'                  ← 確認

CACHES = {                           ← 確認
'default': {
'BACKEND': 'django.core.cache.backends.memcached.MemcachedCache',
'LOCATION': '127.0.0.1:11211',
    }
}
```

```
OPENSTACK_KEYSTONE_DEFAULT_ROLE = "user"   ← 変更
TIME_ZONE = "Asia/Tokyo"
```

次のコマンドを実行して正しく設定を行ったか確認します。

```
controller# less /etc/openstack-dashboard/local_settings.py  | grep -v "^\s*$" | grep -v "^\s*#"
```

念のため、リダイレクトするように設定しておきます（数字は待ち時間）。

```
controller# vi /var/www/html/index.html
...
  <head>
    <meta http-equiv="Content-Type" content="text/html; charset=UTF-8" />
    <meta http-equiv="refresh" content="3; url=/horizon" />    ← 追記
```

変更を反映させるため、Apache とセッションストレージサービスを再起動します。

```
controller# service apache2 restart
```

15.3　Dashboardへのアクセス確認

　controller ノードとネットワーク的に接続されているマシンからブラウザーで以下 URL に接続して OpenStack のログイン画面が表示されるか確認します。

　※ブラウザーで接続するマシンはあらかじめ DNS もしくは/etc/hosts に controller ノードの IP を記述しておくなど、controller ノードの名前解決を行っておく必要があります。

```
http://controller/horizon/
```

　※上記 URL にアクセスしてログイン画面が表示され、ユーザー admin と demo でログイン（パスワード:password）でログインできれば問題ありません。

15.4　セキュリティグループの設定

　OpenStack の上で動かすインスタンスのファイアウォール設定は、セキュリティグループで行います。ログイン後、次の手順でセキュリティグループを設定できます。

1. 対象のユーザーでログイン
2. 「プロジェクト→コンピュート→アクセスとセキュリティ」を選択
3. 「ルールの管理」ボタンをクリック

4. 「ルールの追加」で許可するルールを定義
5. 「追加」ボタンをクリック

セキュリティグループは複数作成できます。インスタンスを起動する際に作成したセキュリティグループを選択することで、セキュリティグループで定義したポートを解放したり、拒否したり、接続できるクライアントを制限できます。

15.5 キーペアの作成

OpenStack ではインスタンスへのアクセスはデフォルトで公開鍵認証方式で行います。次の手順でキーペアを作成できます。

1. 対象のユーザーでログイン
2. 「プロジェクト→コンピュート→アクセスとセキュリティ」をクリック
3. 「キーペア」タブをクリック
4. 「キーペアの作成」ボタンをクリック
5. キーペア名を入力
6. 「キーペアの作成」ボタンをクリック
7. キーペア（拡張子:pem）ファイルをダウンロード

インスタンスに SSH 接続する際は、-i オプションで pem ファイルを指定します。

```
client$ ssh -i mykey.pem cloud-user@instance-floating-ip
```

15.6 インスタンスの起動

前の手順で Glance に CirrOS イメージを登録していますので、早速構築した OpenStack 環境上でインスタンスを起動してみましょう。

1. 対象のユーザーでログイン
2. 「プロジェクト→コンピュート→イメージ」をクリック
3. イメージ一覧から起動する OS イメージを選び、「インスタンスの起動」ボタンをクリック
4. 「インスタンスの起動」詳細タブで起動するインスタンス名、フレーバー、インスタンス数を設定

5. アクセスとセキュリティタブで割り当てるキーペア、セキュリティグループを設定
6. ネットワークタブで割り当てるネットワークを設定
7. 作成後タブで必要に応じてユーザーデータの入力（オプション）
8. 高度な設定タブでパーティションなどの構成を設定（オプション）
9. 右下の「起動」ボタンをクリック

15.7 Floating IPの設定

起動したインスタンスにFloating IPアドレスを設定することで、Dashboardのコンソール以外からインスタンスにアクセスできるようになります。インスタンスにFloating IPを割り当てるには次の手順で行います。

1. 対象のユーザーでログイン
2. 「プロジェクト→コンピュート→インスタンス」をクリック
3. インスタンスの一覧から割り当てるインスタンスをクリック
4. アクションメニューから「Floating IPの割り当て」をクリック
5. 「Floating IP割り当ての管理」画面のIPアドレスで「+」ボタンをクリック
6. 右下の「IPの確保」ボタンをクリック
7. 割り当てるIPアドレスとインスタンスを選択して右下の「割り当て」ボタンをクリック

15.8 インスタンスへのアクセス

Floating IPを割り当てて、かつセキュリティグループの設定を適切に行っていれば、リモートアクセスできるようになります。セキュリティグループでSSHを許可した場合、端末からSSH接続が可能になります（下記は実行例）。

```
client$ ssh -i mykey.pem cloud-user@instance-floating-ip
```

その他、適切なポートを開放してインスタンスへのPingを許可したり、インスタンスでWebサーバーを起動して外部PCからアクセスしてみましょう。

第II部

監視環境 構築編

第16章 Zabbixのインストール

　Zabbix は Zabbix SIA 社が提供するパッケージを使う方法と Canonical Ubuntu が提供するパッケージを使う方法がありますが、今回は新たなリポジトリー追加が不要な Ubuntu が提供する標準パッケージを使って、Zabbix が動作する環境を作っていきましょう。

　なお、Ubuntu の Zabbix 関連のパッケージは universe リポジトリーで管理されています。universe リポジトリーを参照するように/etc/apt/sources.list を設定する必要があります。次のように実行して同じような結果が出力されれば、universe リポジトリーが参照できるように設定されていると判断できます。

```
# apt-cache policy zabbix-server-mysql
zabbix-server-mysql:
  Installed: (none)
  Candidate: 1:2.2.2+dfsg-1ubuntu1
  Version table:
     1:2.2.2+dfsg-1ubuntu1 0
        500 http://us.archive.ubuntu.com/ubuntu/ trusty/universe amd64 Packages
```

　本例では Zabbix を Ubuntu Server 14.04.2 上にオールインワン構成でセットアップする手順を示します。

16.1 パッケージのインストール

　次のコマンドを実行し、Zabbix および Zabbix の稼働に必要となるパッケージ群をインストールします。

第16章　Zabbixのインストール

```
zabbix# apt-get install -y php5-mysql zabbix-agent zabbix-server-mysql \
 zabbix-java-gateway zabbix-frontend-php
```

インストール中にMySQLのパスワードを設定する必要があります。

16.2　Zabbix用データベースの作成

データベースの作成

次のコマンドを実行し、Zabbix用MySQLユーザおよびデータベースを作成します。

```
zabbix# mysql -u root -p << EOF
CREATE DATABASE zabbix CHARACTER SET UTF8;
GRANT ALL PRIVILEGES ON zabbix.* TO 'zabbix'@'localhost' \
  IDENTIFIED BY 'zabbix';
EOF
Enter password:  ← MySQLのrootパスワードを入力 (16-1で設定したもの)
```

次のコマンドを実行し、Zabbix用データベースにテーブル等のデータベースオブジェクトを作成します。

```
zabbix# cd /usr/share/zabbix-server-mysql/
zabbix# zcat schema.sql.gz | mysql zabbix -uzabbix -pzabbix
zabbix# zcat images.sql.gz | mysql zabbix -uzabbix -pzabbix
zabbix# zcat data.sql.gz | mysql zabbix -uzabbix -pzabbix
```

データベースの確認

作成したデータベーステーブルにアクセスしてみましょう。zabbixデータベースに様々なテーブルがあり、参照できれば問題ありません。

```
zabbix# mysql -u root -p
Enter password:           ← パスワードzabbixを入力
mysql> show databases;
+--------------------+
| Database           |
+--------------------+
| information_schema |
| zabbix             |
+--------------------+
2 rows in set (0.00 sec)
mysql> use zabbix;
mysql> show tables;
+----------------------+
| Tables_in_zabbix     |
+----------------------+
```

```
| acknowledges       |
| actions            |
| alerts             |
...

mysql> describe acknowledges;
+----------------+---------------------+------+-----+---------+-------+
| Field          | Type                | Null | Key | Default | Extra |
+----------------+---------------------+------+-----+---------+-------+
| acknowledgeid  | bigint(20) unsigned | NO   | PRI | NULL    |       |
| userid         | bigint(20) unsigned | NO   | MUL | NULL    |       |
| eventid        | bigint(20) unsigned | NO   | MUL | NULL    |       |
| clock          | int(11)             | NO   | MUL | 0       |       |
| message        | varchar(255)        | NO   |     |         |       |
+----------------+---------------------+------+-----+---------+-------+
5 rows in set (0.01 sec)
```

16.3　Zabbix サーバーの設定および起動

/etc/zabbix/zabbix_server.conf を編集し、次の行を追加します。なお、MySQL ユーザ zabbix のパスワードを別の文字列に変更した場合は、該当文字列を指定する必要があります。

```
zabbix# vi /etc/zabbix/zabbix_server.conf
...
DBPassword=zabbix
```

/etc/default/zabbix-server を編集し、起動可能にします。

```
zabbix# vi /etc/default/zabbix-server
...
# Instructions on how to set up the database can be found in
# /usr/share/doc/zabbix-server-mysql/README.Debian
START=yes                    ← no から yes に変更
```

以上の操作を行ったのち、サービス zabbix-server を起動します。

```
zabbix# service zabbix-server restart
```

16.4　Zabbix frontend の設定および起動

PHP の設定を Zabbix が動作するように修正するため、/etc/php5/apache2/php.ini を編集します。

第 16 章　Zabbix のインストール

```
zabbix# vi /etc/php5/apache2/php.ini

[PHP]
...
post_max_size = 16M           ← 変更
max_execution_time = 300      ← 変更
max_input_time = 300          ← 変更

[Date]
date.timezone = Asia/Tokyo    ← 変更
```

Zabbix frontend へアクセスできるよう、設定ファイルをコピーします。

```
zabbix# cp -p /usr/share/doc/zabbix-frontend-php/examples/apache.conf
/etc/apache2/conf-enabled/zabbix.conf
```

これまでの設定変更を反映させるため、サービス Apache2 をリロードします。

```
zabbix# service apache2 reload
```

次に、Zabbix frontend の接続設定を行います。次のコマンドを実行し、一時的に権限を変更します。

```
zabbix# chmod 775 /etc/zabbix
zabbix# chgrp www-data /etc/zabbix
```

Web ブラウザーで Zabbix frontend へアクセスします。画面指示に従い、Zabbix の初期設定を行います。

```
http://<Zabbix frontend の IP アドレス>/zabbix/
```

次のような画面が表示されます。「Next」ボタンをクリックして次に進みます。

- 「2. Check of pre-requisites」は、システム要件を満たしている（すべて OK となっている）ことを確認します。
- 「3. Configure DB connection」は次のように入力し、「Test connection」ボタンを押して OK となることを確認します。

項目	設定値
Database type	MySQL
Database host	localhost
Database Port	0
Database name	zabbix
User	zabbix
Password	zabbix

16.4 Zabbix frontend の設定および起動

図 16.1　Zabbix 初期セットアップ

- 「4. Zabbix server details」は Zabbix Server のインストール場所の指定です。本例ではそのまま次に進みます。
- 「5. Pre-Installation summary」で設定を確認し、問題なければ次に進みます。
- 「6. Install」で設定ファイルのパスが表示されるので確認し「Finish」ボタンをクリックします（/etc/zabbix/zabbix.conf.php）。
- ログイン画面が表示されるので、Admin/zabbix（初期パスワード）でログインします。

Zabbix の初期セットアップ終了後にログイン画面が表示されますので、実際に運用開始する前に次のコマンドを実行して権限を元に戻します。

```
zabbix# chmod 755 /etc/zabbix
zabbix# chgrp root /etc/zabbix
```

第17章 Hatoholのインストール

　Hatohol は CentOS6.5 以降、Ubuntu Server 12.04 および 14.04 などで動作します。CentOS6.5 以降および 7.x 向けには導入に便利な RPM パッケージが公式で提供されています。

　本例では Hatohol を CentOS 7 上にオールインワン構成でセットアップする手順を示します。

図 17.1　Hatohol ダッシュボード

第 17 章 Hatohol のインストール

17.1 インストール

1. Hatohol をインストールするために、Project Hatohol 公式の YUM リポジトリーを登録します。

```
hatohol# wget -P /etc/yum.repos.d/ http://project-hatohol.github.io/repo/hatohol-el7.repo
```

2. EPEL リポジトリー上のパッケージのインストールをサポートするため、EPEL パッケージを追加インストールします。

```
hatohol# yum install -y epel-release
hatohol# yum update
```

3. Hatohol サーバーをインストールします。

```
hatohol# yum -y install hatohol-server
```

4. Hatohol Web Frontend をインストールします。

```
hatohol# yum install -y hatohol-web
```

5. 必要となる追加パッケージをインストールします。

```
hatohol# yum install -y mariadb-server qpid-cpp-server
```

17.2 セットアップ

1. /etc/my.cnf の編集

- 編集前

```
!includedir /etc/my.cnf.d
```

- 編集後

```
includedir /etc/my.cnf.d
```

2. /etc/my.cnf.d/server.cnf の編集

セクション [mysqld] に、少なくとも次のパラメーターを追記します。

```
[mysqld]
character-set-server = utf8
skip-character-set-client-handshake
default-storage-engine = innodb
innodb_file_per_table
```

3. MariaDB サービスの自動起動の有効化と起動

```
hatohol# systemctl enable mariadb.service
hatohol# systemctl start mariadb.service
```

4. MariaDB ユーザ root のパスワード変更

```
hatohol# mysqladmin password
```

5. Hatohol DB の初期化

```
hatohol# hatohol-db-initiator --db_user root --db_password <4 で設定した root パスワード>
...
Succeessfully loaded: /usr/bin/../share/hatohol/sql/init-user.sql
Succeessfully loaded: /usr/bin/../share/hatohol/sql/server-type-zabbix.sql
Succeessfully loaded: /usr/bin/../share/hatohol/sql/server-type-nagios.sql
Succeessfully loaded: /usr/bin/../share/hatohol/sql/server-type-hapi-zabbix.sql
Succeessfully loaded: /usr/bin/../share/hatohol/sql/server-type-hapi-json.sql
Succeessfully loaded: /usr/bin/../share/hatohol/sql/server-type-ceilometer.sql
```

初期状態で上記コマンドを実行した場合、MySQL ユーザ hatohol、データベース hatohol が作成されます。これらを変更する場合、事前に/etc/hatohol/hatohol.conf を編集してください。

6. Hatohol Web 用 DB の作成

```
hatohol# mysql -u root -p << EOF
CREATE DATABASE hatohol_client;
GRANT ALL PRIVILEGES ON hatohol_client.* TO 'hatohol'@'localhost' \
  IDENTIFIED BY 'hatohol';
EOF
```

7. Hatohol Web 用 DB へのテーブル追加

```
hatohol# /usr/libexec/hatohol/client/manage.py syncdb
```

8. Hatohol サーバーの自動起動の有効化と起動

```
hatohol# systemctl enable hatohol.service
hatohol# systemctl start hatohol.service
```

9. Hatohol Web の自動起動の有効化と起動

第 17 章　Hatohol のインストール

```
hatohol# systemctl enable httpd.service
hatohol# systemctl start httpd.service
```

10. Hatohol および Apache Web サーバーの動作確認

```
hatohol# systemctl status -l hatohol.service
hatohol# systemctl status -l httpd.service
```

17.3　セキュリティ設定の変更

　CentOS インストール後の初期状態では、SElinux、Firewalld、iptables といったセキュリティ機構により他のコンピュータからのアクセスに制限が加えられます。Hatohol を使用するにあたり、これらを適切に解除する必要があります。

1. SELinux の設定

```
hatohol# getenforce
Enforcing
```

　Enforcing の場合、次のコマンドで SElinux ポリシールールの強制適用を解除できます。

```
hatohol# setenforce 0
hatohol# getenforce
Permissive
```

　恒久的に SELinux ポリシールールの適用を無効化するには、/etc/selinux/config を編集します。

- 編集前

```
SELINUX=enforcing
```

- 編集後

```
SELINUX=permissive
```

完全に SELinux を無効化するには、次のように設定します。

```
SELINUX=disabled
```

> 筆者注:
> SELinux はできる限り無効化すべきではありません。

2. パケットフィルタリングの設定フィルタリングの設定変更は、次のコマンドで恒久的に変更可能です。

```
hatohol# firewall-cmd --add-service=http --zone=public
hatohol# iptables-save > /etc/sysconfig/iptables
```

17.4 Hatoholによる情報の閲覧

Hatohol Web が動作しているホストのトップディレクトリーを Web ブラウザーで表示してください。10.0.0.10 で動作している場合は、次の URL となります。admin/hatohol（初期パスワード）でログインできます。

```
http://10.0.0.10/
```

Hatohol は監視サーバーから取得したログ、イベント、性能情報を表示するだけでなく、それらの情報を統計してグラフとして出力できる機能が備わっています。CPU のシステム時間、ユーザー時間をグラフとして出力すると次のようになります。

17.5 HatoholにZabbixサーバーを登録

Hatohol をインストールできたら、Zabbix サーバーの情報を追加します。Hatohol Web にログインしたら、上部のメニューバーの「設定→監視サーバー」をクリックします。「監視サーバー」の画面に切り替わったら「監視サーバー追加」ボタンをクリックしてノードを登録します。

項目	設定値
監視サーバータイプ	Zabbix
ニックネーム	zabbix1
ホスト名	zabbix
IP アドレス	(Zabbix サーバーの IP アドレス)
ポート番号	80
ユーザー	Admin
パスワード	zabbix

ページを再読み込みして、通信状態が「初期状態」から「正常」になることを確認します。

第 17 章　Hatohol のインストール

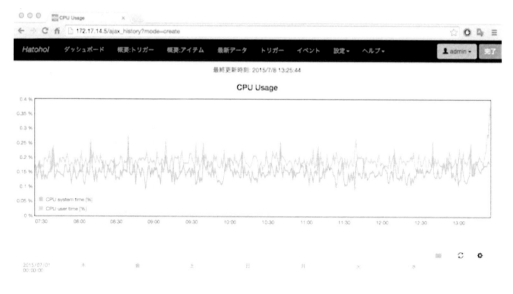

図 17.2　Hatohol のグラフ機能

図 17.3　Zabbix サーバーの追加

17.6　HatoholでZabbixサーバーの監視

　インストール直後の Zabbix サーバーはモニタリング設定が無効化されています。これを有効化すると Zabbix サーバー自身の監視データを取得できるようになり、Hatohol で閲覧できるよ

うになります。

Zabbix サーバーのモニタリング設定を変更するには、次の手順で行います。

- Zabbix のメインメニュー「Configuration → Host groups」をクリックします。
- Host groups 一覧から「Zabbix server」をクリックします。
- 「Zabbix server」の Host の設定で、Status を「Monitored」に変更します。
- 「Save」ボタンをクリックして設定変更を適用します。

以上の手順で、Zabbix サーバーを監視対象として設定できます。

17.7　Hatohol でその他のホストの監視

　Zabbix と Hatohol の連携ができたので、あとは対象のサーバーに Zabbix Agent をインストールし、手動で Zabbix サーバーにホストを追加するか、ディスカバリ自動登録を使って、特定のネットワークセグメントに所属する Zabbix Agent がインストールされたホストを自動登録するようにセットアップするなどの方法で監視ノードを追加できます。追加したノードは Zabbix および Hatohol で監視できます。

Zabbix Agent のインストール

　Zabbix で OpenStack の controller ノード、network ノード、compute ノードを監視するために Zabbix Agent をインストールします。Ubuntu には標準で Zabbix Agent パッケージが用意されているので、apt-get コマンドなどを使ってインストールします。

```
# apt-get update && apt-get install -y zabbix-agent
```

Zabbix Agent の設定

　Zabbix Agent をインストールしたら次にどの Zabbix サーバーと通信するのか設定を行う必要があります。最低限必要な設定は次の 3 つです。次のように設定します。

　(controller ノードの設定記述例)

```
# vi /etc/zabbix/zabbix_agentd.conf
...
Server          10.0.0.10      ← Zabbix サーバーの IP アドレスに書き換え
ServerActive    10.0.0.10      ← Zabbix サーバーの IP アドレスに書き換え
Hostname        controller     ← Zabbix サーバーに登録する際のホスト名と同一のものを設定
ListenIP        10.0.0.101     ← Zabbix エージェントが待ち受ける側の IP アドレス
```

第 17 章　Hatohol のインストール

　ListenIP に指定するのは Zabbix サーバーと通信できる NIC に設定した IP アドレスを設定します。

　変更した Zabbix Agent の設定を反映させるため、Zabbix Agent サービスを再起動します。

```
# service zabbix-agent restart
```

ホストの登録

　Zabbix Agent のセットアップが終わったら、次に Zabbix Agent をセットアップしたサーバーを Zabbix の管理対象として追加します。次のように設定します。

- 「Configuration → Host」をクリックします。初期設定時は Zabbix server のみが登録されていると思います。同じように監視対象サーバーを Zabbix に登録します。
- 「Hosts」画面の右上にある、「Create Host」ボタンをクリックします。
- 次のように設定します。

「Host」の設定	説明
Host name	zabbix_agentd.conf にそれぞれ記述した Hostname を記述
Visible name	表示名（オプション）
Groups	所属グループの指定。例として Linux servers を指定
Agent interfaces	監視対象とする Agent がインストールされたホストの IP アドレス（もしくはホスト名）
Status	Monitored

　その他の項目は適宜設定します。

- 「CONFIGURATION OF HOSTS」の「Templates」タブをクリックして設定を切り替えます。
- 「Link new templates」の検索ボックスに「Template OS Linux」と入力し、選択肢が出てきたらクリックします。そのほかのテンプレートを割り当てるにはテンプレートを検索し、該当のものを選択します。
- 「Link new templates」にテンプレートを追加したら、その項目の「Add」リンクをクリックします。「Linked templates」に追加されます。
- 「Save」ボタンをクリックします。
- 「Hosts」画面にサーバーが追加されます。ページの再読み込みを実行して、Zabbix エージェントが有効になっていることを確認してください。「Z」アイコンが緑色になれば OK

です。

図 17.4　Zabbix エージェントステータスを確認

- ほかに追加したいサーバーがあれば「Zabbix Agent のインストール、設定、ホストの登録」の一連の流れを繰り返します。監視したい対象が大量にある場合はオートディスカバリを検討してください。

Hatohol で確認

登録したサーバーの情報が Hatohol で閲覧できるか確認してみましょう。Zabbix サーバー以外のログなど表示できるようになれば OK です。

参考情報

ホストの追加やディスカバリ自動登録については次のドキュメントをご覧ください。

- https://www.zabbix.com/documentation/2.2/jp/manual/quickstart/host
- https://www.zabbix.com/documentation/2.2/jp/manual/discovery/auto_registration
- http://www.zabbix.com/jp/auto_discovery.php
- https://www.zabbix.com/documentation/2.2/jp/manual/discovery/network_discovery/rule

17.8　Hatohol Arm Plugin Interface を使用する場合の操作

Hatohol Arm Plugin Interface(HAPI) を使用する場合、/etc/qpid/qpidd.conf に次の行を追

第 17 章　Hatohol のインストール

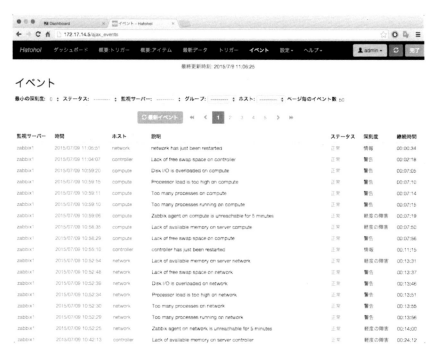

図 17.5　OpenStack ノードの監視

記します。なお、=の前後にスペースを入れてはなりません。

付録A　FAQフォーラム参加特典について

本書の購入者限定特典としてFAQフォーラム（Googleグループ）を用意しています。Googleアカウントにログインのうえ下記URLにアクセスし、本フォーラムの「メンバー登録を申し込む」をクリックしてください。申し込みの際に追加情報として購入した書籍名をご入力ください。

https://groups.google.com/d/forum/vtj-openstack-faq

本フォーラムは書籍の購入者限定のサービスです。フォーラム内のすべての質問に対して日本仮想化技術の社員が回答を行うことをお約束するものではないこと、サービスの公開期間は書籍出版後2年間（〜2018/03）を予定していること、前記期間内であってもOpenStackおよび関連サービスの仕様変更に伴いフォーラムを継続できなくなる可能性があることをご了承ください。また、本書およびフォーラムの解説内容によって生じる直接または間接被害について、著者である日本仮想化技術株式会社ならびに株式会社インプレスでは一切の責任を負いかねます。

●著者紹介

遠山 洋平
日本仮想化技術株式会社
1981 年 6 月、宮城県生まれ。2008 年に日本仮想化技術株式会社に入社し、仮想化技術の検証、ベンチマークおよび構築などに従事。Linux やサーバ仮想化、デスクトップ仮想化に関連した記事を雑誌や書籍、Web などに多数執筆。社内ではデスクトップ仮想化や OpenStack、Docker や LXC などの構築や検証を担当。

●スタッフ
- 田中 佑佳（表紙デザイン）
- 鈴木 教之（編集、紙面レイアウト）

本書のご感想をぜひお寄せください
http://book.impress.co.jp/books/1115101150
アンケート回答者の中から、抽選で商品券（1万円分）や図書カード（1,000円分）などを毎月プレゼント。
当選は商品の発送をもって代えさせていただきます。

●本書の内容に関するご質問は、書名・ISBN・お名前・電話番号と、該当するページや具体的な質問内容、お使いの動作環境などを明記のうえ、インプレスカスタマーセンターまでメールまたは封書にてお問い合わせください。電話やFAX等でのご質問には対応しておりません。なお、本書の範囲を超える質問に関しましてはお答えできませんのでご了承ください。

●落丁・乱丁本はお手数ですがインプレスカスタマーセンターまでお送りください。送料弊社負担にてお取り替えさせていただきます。但し、古書店で購入されたものについてはお取り替えできません。

■読者の窓口
インプレスカスタマーセンター
〒101-0051 東京都千代田区神田神保町一丁目105番地
TEL 03-6837-5016 ／ FAX 03-6837-5023
info@impress.co.jp

■書店／販売店のご注文窓口
株式会社インプレス 受注センター
TEL 048-449-8040
FAX 048-449-8041

OpenStack構築手順書 Kilo版（Think IT Books）

2016年5月1日 初版発行

著　者　日本仮想化技術株式会社
発行人　土田 米一
編集人　高橋 隆志
発行所　株式会社インプレス
　　　　〒101-0051　東京都千代田区神田神保町一丁目105番地
　　　　TEL　03-6837-4635（出版営業統括部）
　　　　ホームページ　http://book.impress.co.jp/

本書は著作権法上の保護を受けています。本書の一部あるいは全部について（ソフトウェア及びプログラムを含む）、株式会社インプレスから文書による許諾を得ずに、いかなる方法においても無断で複写、複製することは禁じられています。

Copyright © 2016 Virtual Tech JP. All rights reserved.
印刷所　京葉流通倉庫株式会社
ISBN978-4-8443-8057-3　C3055
Printed in Japan